Principles of Random Variate Generation

Principles of Random Variate Generation

John Dagpunar

Department of Mathematics and Computer Studies
Dundee College of Technology

CLARENDON PRESS · OXFORD
1988

Oxford University Press, Walton Street, Oxford OX2 6DP
Oxford New York Toronto
Delhi Bombay Calcutta Madras Karachi
Petaling Jaya Singapore Hong Kong Tokyo
Nairobi Dar es Salaam Cape Town
Melbourne Auckland
and associated companies in
Berlin Ibadan

Oxford is a trade mark of Oxford University Press

Published in the United States
by Oxford University Press, New York

British Library Cataloguing-in-Publication data:
Dagpunar, John
Principles of random variate generation.
1. Random variables
I. Title.
519.2
ISBN 0-19-852202-9

Library of Congress Cataloging-in-Publication Data
Dagpunar, John.
Principles of random variate generation/John Dagpunar.
Bibliography:
Includes index.
1. Random variables.   I. Title.   II. Title: Random variate
generation
QA273.D24   1988   519.2—dc19   88-3212
ISBN 0-19-852202-9

Typeset by Macmillan India Ltd, Bangalore 25
Printed in Great Britain
at the University Printing House, Oxford
by David Stanford
Printer to the University

Preface

This book is concerned with the theory and practice of 'generating' or 'sampling' observations from specified probability distributions. This task is known as random variate generation, to distinguish it from random number generation, which we take to be the sampling of observations uniformly distributed in the interval [0, 1).

The book is intended for a fairly wide readership. Increasingly statisticians are coming to realize that simulation can, when used carefully, prove to be a powerful and respectable numerical tool, where analytical methods have failed. There is a vast literature on random variate and number generation, but it is widely dispersed, which raises a problem for the busy statistician, who needs to simulate some process as part of a larger investigation. One objective of the book, therefore, is to consolidate this material. It is hoped that the book will appear on a professional statistician's bookshelf, providing a reference work, giving both theory and practical guidance as to the best generation method to use in particular circumstances. The idea for the book arose out of a course of lectures I gave on the Statistical Aspects of Simulation in the Department of Mathematics and Statistics at Brunel University. It is anticipated that the book could be used for similar courses at advanced undergraduate or graduate level. A third category of reader will be those who use simulation, whether it be in operational research, the physical or biological sciences, computer science, or engineering. Experience of the first area has suggested to me that the quality of the models would be much improved by paying more attention to the statistical aspects, one of which is random variate and number generation.

Chapter 1 gives a brief outline of the nature and history of simulation, discussing the requirement for random variates, and how the efficiency of such generators can be measured. All random variates begin life as humble random numbers, and it is essential that a good source of these is available. The methods available are discussed in Chapter 2. Although the linear congruential generator is almost universally used today, its limitations are severe. In the future we may expect Tauseworthe generators, which have the potential for a much greater degree of multi-dimensional uniformity, to be used, once their properties are better understood. Chapter 2 also deals with a selection of empirical tests of randomness, together with the powerful theoretical tests related to lattice behaviour for the linear congruential methods.

The remainder of the book deals with the transformation of random numbers to obtain the required random variates. A wide set of principles has emerged, and these are described, largely without reference to particular distributions, in Chapter 3. Chapters 4 and 5 provide methods of generation from specific continuous and discrete distributions. For completeness, most of the methods available for any given distribution are described. Using theoretical measures of performance, and in some cases computer timings, recommendations for particular distributions are given. In many cases these recommendations depend upon the application (for example, whether the parameter values of a distribution change between calls of a generator). Algorithms are provided for those readers wishing to program their own routines. Multivariate distributions, including some reference to problems arising in a reliability context are discussed in Chapter 6. Chapter 7 brings together some miscellaneous topics, including the generation of order statistics, the heterogeneous Poisson process, the efficient simulation of Markov chains and processes, and tail variate generation. The book also provides an opportunity to describe some as yet unpublished work of mine, including the generation of two-parameter zeta variates, of generalized inverse Gaussian variates using a ratio method, and of tail variates from the t and normal distributions, and the simulation of a fission spectrum arising in nuclear physics.

Finally, what of computing? The last ten years have seen a movement away from the use of large impersonal computers to the micro-computer. Software support is often not as comprehensive. For this reason it was felt it would be advantageous to include a set of short (FORTRAN 77) subroutines for the major distributions. These appear as Appendix 1. In spite of the computer revolution, I have still found it necessary to be able to generate small samples (under 20 observations) manually from specific distributions. For this reason I have described a graphical method (Appendix 2) for the manual generation of gamma and normal random variates. These variates may themselves be used as a basis for sampling from beta, t, and chi-squared distributions.

Several people helped make this book possible. In particular I would like to thank Dr I. D. Hill of the MRC Clinical Research Centre, Harrow, who kindly read the original manuscript, suggesting several improvements, including some major revisions to the FORTRAN routines. Any remaining deficiencies are of course my responsibility. I am also indebted to the staff of Oxford University Press for supporting the project, and to Mrs. S. Nicoll and Mrs J. Edmonds, who typed the manuscript.

Dundee   J. S. D.
June 1987

Contents

Contents

To Bridget, Paul, Stephen,
and Rebecca

Algorithmic conventions

To encourage readers to program methods in their own preferred languages, algorithms are given when appropriate. The algorithmic (pseudo-code) conventions employed include the following:

Input [n, p] The algorithm requires input values for the variables n and p. The input list may include user provided functions.

Output [X] The algorithm outputs a value for the variable X.

Saved [a, b, c] The saved variables a, b, c are those whose values are assigned only when the input values of n and p have changed since the last call of the algorithm. The values of a, b, and c are retained between successive calls of the algorithm.

Parameter [$d = 5.0$] The constant d is assigned a value of 5.0 on the first call of the algorithm.

Line numbering An algorithm consists of a sequence of numbered lines. The calculation of saved variable values may be thought of as a 'set-up' calculation contained in lines labelled S1, S2, . . ., the remaining lines being labelled 1, 2, ... Thus S1, S2, . . . are executed on the first call of the algorithm, or if the input parameter values have changed since the last call. Otherwise S1, S2, . . . are bypassed and execution begins at line 1.

:= 'Becomes equal to'. The value of the expression to the right of := is assigned to the variable or parameter to the left of := .

= 'Is equal to'. The value of the expression to the right of = is mathematically equal to the value of the expression to the left of = .

Print x, y Print the values of x and y.

Exit Leave the algorithm with current output and saved variable values.

Stop   Execution is halted, usually because an input parameter value is outside a specified range.

Data: $a/2/$, $b/5/$ User specified initial values of 2 and 5 for variables a and b, respectively.

For $j := n, m$ 'For–Next loop'. Set $j := n$ and execute $e_1, e_2, \ldots$. Incre-
   e_1 ment j by 1 and repeat, finishing with $j := \text{Max}(n, m)$. Thus
   e_2 if $m < n$ no statements are executed.
   $\vdots$
Next j

While e_1 do   'Do–While loop'. If e_1 is true execute $e_2, e_3, \ldots$. Repeat
   e_2 while e_1 remains true.
   e_3
   $\vdots$
End While

Repeat   'Repeat–Until loop'. Execute $e_1, e_2, \ldots$ and repeat until
   e_1 e_0 is true.
   e_2
   $\vdots$
Until e_0

If e_1 goto n If e_1 is true goto statement labelled n.
If e_1 exit   If e_1 is true exit.
If e_1 stop   If e_1 is true stop.

If e_1 then   'If–then–end if block'. If e_1 is true execute $e_2, e_3, \ldots$.
   e_2
   e_3
   $\vdots$
End If

If e_1 then   'If–then–else–end if block'. If e_1 is true execute $e_2, e_3, \ldots$,
   e_2 otherwise execute $f_1, f_2, \ldots$.
   e_3
   $\vdots$
Else
   f_1
   f_2
   $\vdots$
End If

If e_1 then   'If–then–else if–end if block'. If e_1 is true execute e_2,
  e_2 $e_3, \ldots$, otherwise if g_1 is true execute $f_1, f_2, \ldots$
  e_3
  $\vdots$

Else if g_1 then
  f_1
  f_2
  $\vdots$
End If

Glossary

$U(0, 1)$	Uniformly distributed in the interval $(0, 1)$.
$N(0, 1)$	Normally distributed with zero mean, and unit variance.
R	A random number $\sim U[0, 1)$.
p.d.f.	Probability density function.
p.m.f.	Probability mass function.
c.d.f.	Cumulative distribution function.
i.i.d.	Identically and independently distributed.
$E(X)$	Expectation of a random variable, X.
$V(X)$	Variance of a random variable, X.
$\rho(X, Y)$	Correlation between X and Y.
$X_{(j)}$	The jth order statistic.
$\Gamma(\alpha)$	The gamma function $(\alpha > 0)$.
$B(\alpha, \beta)$	The beta function $(\alpha > 0, \beta > 0)$.
$\ln x$	Natural logarithm of x.
$\langle x \rangle$	The largest integer not exceeding x.
$\lvert x \rvert$	The absolute value of x.
x', A'	The transposes of vector x, matrix A.
$\lvert x \rvert$	The length of vector x.
LCM	Lowest common multiple.
GCD	Greatest common divisor.
$A \cup B$	The union of sets A and B.
$A \cap B$	The intersection of sets A and B.

1
Simulation and random variate generation

With the development of computers, simulation is now a frequently used method for solving problems in engineering, the physical sciences, economics, and the behavioural sciences. Almost all simulations need a source of observations drawn from specified probability distributions. For example, a particular simulation may include a requirement for a random sample of 10 000 observations drawn from a normal distribution with specified mean and variance. This task is known as *random variate generation* and it is the main subject matter of this book. In order to set this activity in context, this first chapter gives a brief description of the nature of simulation.

1.1 The nature and origins of simulation

Any scientist knows that there are some problems which are not amenable to a totally mathematical solution. For example, a production engineer may wish to model the behaviour of a manufacturing process consisting of interactions between men, machinery, and raw materials over time. In the simplest of cases the system might reduce to a queueing problem whose behaviour can be represented by a set of differential difference equations. If he is extremely fortunate the engineer may be able to solve these equations analytically or numerically. In a more complex system, it may still be possible to formulate the behaviour of the system as a set of equations, but it may prove impossible or be deemed impracticable to solve these either analytically or numerically. In the third and most complicated of situations it may not even be possible to describe the behaviour of the system through a set of equations.

Faced with the second and third situations, the engineer might consider trying to understand the behaviour of the system by conducting experiments on it under real-life conditions. The disadvantages of such an approach are obvious. It will probably be more time consuming than modelling on paper or computer, it may interrupt the smooth functioning of the plant, it is likely to incur more expense, and might even be dangerous. An alternative strategy is to map the real-life system onto a *model*, upon which sampling experiments are carried out. In these experiments events are generated in some specified probabilistic fashion. Statistics may then be constructed from these experiments. If the properties of the statistics are relevant to the measures of interest in the real-life system, then some inference can be made about the behaviour of the system. This approach is known as *simulation*. The success of such a method will depend upon how rapidly events can be generated within the simulation, and the degree of association between the statistics and the measures of performance of the system itself.

The characteristic feature of most simulation experiments is the sampling or generation of probabilistic events. It is understandable, therefore, that it is often thought that simulation is used only for solving problems in which there are some probabilistic elements. Historically, some of the first simulations were performed in the 1940s under the name of *Monte Carlo* methods. These experiments were used to solve deterministic problems arising within theoretical physics. These problems were concerned with the evaluation of multi-dimensional integrals and solution of partial differential and integral equations.

The distinctive feature of many of these Monte Carlo experiments was that a deterministic problem was mapped onto a sampling experiment. A random sample was collected from the experiment, and the known statistical properties of the statistic proved useful in estimating the solution to the original problem. The sampling experiment required a source of *random numbers*. For the moment we may think of these as a sequence of independent observations of a random variable uniformly distributed in the interval $[0, 1)$. Thus one definition of a Monte Carlo experiment is any technique which utilizes random numbers. Later the techniques proved to be of much wider applicability, and were used for the study of complex probabilistic systems evolving over time. The term simulation is now commonly used to describe all applications.

The Monte Carlo principle may be illustrated through the following problem. Suppose it is desired to evaluate the integral

$$I_\alpha = \int_0^\infty x^{\alpha-1} e^{-x} dx \tag{1.1}$$

for any positive real value of α. We will suppose (somewhat artificially) that there is no access to tables of the gamma function (which will have been constructed by numerical integration procedures), that integration by parts yields $I_\alpha = (\alpha - 1)I_{\alpha-1}$, and that $I_1 = 1$. It is necessary therefore only to evaluate I_α for non-integral values of α in the range $(1, 2)$.

One approach is to consider a random variable X, which is negative exponentially distributed with unit mean. This has probability density function (p.d.f.)

$$f_X(x) = e^{-x} \qquad (x \geq 0).$$

From eqn (1.1) an alternative representation for I_α is $I_\alpha = E(X^{\alpha-1})$. To *estimate* I_α, draw a random sample $X_1, X_2, \ldots, X_n$ and set

$$\hat{I}_\alpha = \frac{1}{n} \sum_{i=1}^{n} X_i^{\alpha-1}.$$

Clearly $E(\hat{I}_\alpha) = I_\alpha$ and the variance, $V(\hat{I}_\alpha) = V(X^{\alpha-1})/n$. Thus the statistic $\hat{I}_\alpha$ constructed from the sampling experiment provides an unbiased estimator of I_α. Table 1.1 shows values for $\hat{I}_\alpha$ as n increases in a particular experiment when $\alpha = 1.375$. Also shown in the table is an alternative estimator $\hat{J}_\alpha$, which we will describe shortly.

Table 1.1 Estimates for I_α with increasing sample size ($\alpha = 1.375$)

n	1	2	3	4	5	6	7	8	9	10
$\hat{I}_\alpha$	1.041	0.638	0.571	0.742	0.923	1.031	1.070	1.045	1.007	0.965
$\hat{J}_\alpha$	0.875	0.911	0.907	0.902	0.911	0.914	0.911	0.906	0.902	0.900

The known value for I_α is 0.888 91 (to five decimal places) and $\hat{I}_\alpha$ approaches this value (slowly) with increasing n. Since the variance is inversely proportional to sample size the reliability of our estimate will be proportional to the square root of the amount of computational effort we are willing to expend. We may quantify the reliability or precision by writing

$$V(\hat{I}_\alpha) = \frac{1}{n} [E(X^{2\alpha-2}) - \{E(X^{\alpha-1})\}^2]$$

$$= \frac{1}{n} \{I_{2\alpha-1} - I_\alpha^2\}.$$

The purpose of the experiment was to estimate I_α, so I_α and $I_{2\alpha-1}$ are unknown. We may either approximate by inserting values estimated from two simulation experiments or 'cheat' by inserting known values, obtained from tables of the gamma function. The latter gives

$$V(\hat{I}_\alpha) = \frac{1}{n} [\Gamma(1.75) - \{\Gamma(1.375)\}^2] = \frac{0.1289}{n},$$

indicating that something of the order of 12 million observations are required for three-figure accuracy! This large expenditure of computer effort is a general feature of many simulation experiments and provides a good reason why simulation should not be used unless analytical or conventional numerical methods fail to give sufficiently accurate answers within a reasonable amount of time. It also provides a motivation for ensuring a good experimental design, in an attempt to reduce the variance, and hence increase the precision. Fortunately the simulator is in a privileged position, since he is conducting the experiments on a computer, and accordingly has a high degree of control over the environment.

In an attempt to increase the precision, consider the random variables X and Y having joint distribution function

$$F_{X,Y}(x, y) = \begin{cases} 1-e^{-x}-e^{-y} & (e^{-x}+e^{-y} \leqslant 1) \\ 0 & \text{(elsewhere)}. \end{cases}$$

The marginal p.d.f.s of X and Y are again negative exponential, mean one. However, X and Y are negatively correlated. This latter property can be utilized to reduce the variance of the estimator. In our revised experiment, we now draw a random sample (X_1, Y_1), (X_2, Y_2), ..., (X_n, Y_n). A new estimator for I_α is

$$\hat{J}_\alpha = \frac{1}{2n} \sum_{i=1}^{n} (X_i^{\alpha-1} + Y_i^{\alpha-1}).$$

This is unbiased and has variance

$$\frac{1}{4n}\{V(X^{\alpha-1}) + V(Y^{\alpha-1}) + 2\,\mathrm{cov}(X^{\alpha-1}, Y^{\alpha-1})\}.$$

The negativity of the last term in this expression ensures that the variance is smaller than would have been the case if X and Y were uncorrelated. In fact, when $\alpha = 1.375$ numerical methods show the variance of $\hat{J}_\alpha$ to be approximately $0.0009/n$. This represents an approximately 140-fold reduction in variance. As a result the sample size may be reduced by a factor of 140, while maintaining the same precision as that obtained with $\hat{I}_\alpha$. Such methods are called *variance reduction* techniques. The final row of Table 1.1 shows values of $\hat{J}_\alpha$, using the same random numbers, and appears to indicate that the true value of I_α is now being approached more rapidly, as a result of the reduced variance of the estimator. Of course a fair comparison of the experiments would take account of the doubling of effort in the second experiment, so even if there was zero correlation the variance would have been reduced by a factor of $1/2$.

1.2   Simulation of dynamic, stochastic systems

The present applications of simulation fall less in the early Monte Carlo area and more in the field of tangible physical systems, often involving random processes evolving over time. Such systems are *stochastic* (rather than deterministic) and *dynamic* (rather than static). This shift in application and emphasis is due partly to the fact that conventional numerical methods have in general achieved more success in solving purely deterministic mathematical problems. It is also due to the emergence of disciplines such as operational research and systems engineering, together with the increasing complexity of systems now being studied in the physical sciences.

It is not the purpose of this book to describe detailed methods for programming such simulation models. However, it will be useful to see how a modest simulation might be structured and note its requirement for random variates. Consider a national lending library, which lends books out to other libraries when requested to do so. We want to determine how the service will vary according to the number of copies of a particular book held by the central library. If all copies happen to be out on loan when a request arrives, the request is put into a queue. It is reasonable to assume that service will be measured by the waiting time which elapses between the arrival and fulfilment of a request. Thus the required output from such a simulation will be the waiting times of successive prospective readers or 'customers'. This output might be stored on disk or tape or be printed out for visual inspection. Note that the output observations form a time series and are not statistically independent. Ultimately the output observations would be analysed to estimate, for example, the proportion of customers who do not have to wait, the mean waiting time of customers, the probability that a customer has to wait more than a specified time, and perhaps the correlation in waiting times between any two customers.

We will suppose that data analysis of requests indicates that the request arrival pattern is essentially a time-dependent Poisson process, reflecting the fact that customers are drawn from a large population, and tend to act independently, but that certain times of the year or week give rise to greater or smaller than average levels of activity. The duration of loan depends on the policies administered by the library and the habits of borrowers. A distribution of loan duration may be hypothesized, tested against available data, and if satisfactory used in the simulation.

At any point in the simulation it is necessary to know how much time has elapsed in the system (the simulation or clock time), the due times at which each book currently on loan is due back, the due time for the next request, and the times of arrival of each of the requests currently in the queue. The latter are stored in a table or array. They are needed so that when a request is satisfied the clock time may be compared with the arrival time to give the waiting time.

Any simulation program needs a time-advance mechanism. In this case the *state* of the system can change only at a request arrival or when a copy is returned. By finding the minimum of the time of the next arrival and the due return times for all books on loan, the time of the next state change may be determined. Figure 1.1 shows how the clock time is advanced from the current position A to B, the time of the next state change.

This way of advancing time in a simulation program may be termed the *time to next event* method. It is an effective and efficient method, since the system is observed at event times only, and no information is lost. An alternative method is to advance time in constant increments. If this is done the time increments have to be small enough to ensure that the chance of two or more state changes is small. Unless events are known to be synchronized with the clock movements this method suffers from inefficiency (many time increments will involve no state changes) and from a possible loss of information. Most efficient simulations use the time to next event method.

When time has been advanced to the next event, the state of the system is changed accordingly. In Fig. 1.1 it would indicate that the third copy of the book has just been returned. If the array of queue request arrival times is empty, then the due time of return of the third copy could be set to ∞, reflecting the fact that the book is now available on the shelf. Otherwise, the waiting time for the customer at the head of the queue is computed (by comparing clock time with request arrival time) and printed out. The request is removed from the queue, with all remaining requests moving up one space in the array. Finally a new due time for copy 3 to be returned is computed by adding the clock time to a loan duration time, which is sampled or generated from the specified distribution.

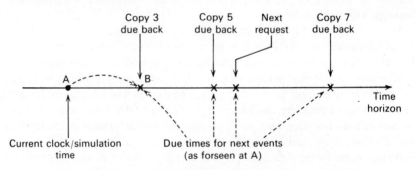

Fig. 1.1. Time advance mechanism using time to next event

The simulation model demonstates the requirement for random variates. One of these is a non-negative loan duration S sampled from a specified p.d.f. $f_S(\cdot)$. The other is a sequence of inter-request times in a time-dependent Poisson process, in which the number of events $N(t)$ occurring in time t is specified by

$$\text{Prob}\{N(t) = n\} = \frac{1}{n!}\left[\exp\left\{-\int_0^t \lambda(u)\mathrm{d}u\right\}\left\{\int_0^t \lambda(u)\mathrm{d}u\right\}^n\right]$$

where $n = 0, 1, \ldots$ and $\lambda(\cdot)$ is the rate of the process. We will see in Chapter 7 that there are several techniques available for generating events from such a process.

The simulation previously outlined is of the *discrete event* type, rather than *continuous*. In the former case the state of the system changes only at discrete time instants, while in the latter the state may change continuously with time, as for example in the simulation of a reservoir system. One important difference between the library simulation and the early Monte Carlo experiments is that the system is *dynamic*, that is the behaviour changes over time. This usually means that output observations are not statistically independent. In the library system, and in most problems involving some queueing element, there is a positive correlation between the output observations. This is known as *autocorrelation*. The more highly congested a queueing system is, the greater will be the autocorrelation. Unfortunately this means that the variance of any statistics we construct will be larger than would be the case if the observations were uncorrelated. This in turn implies that run-lengths have to be correspondingly longer in order to obtain a required precision in the estimates. The autocorrelation also means that the methods used for estimation have to be a little more sophisticated. One method available is to replicate the experiment obtaining independent statistic values, which can then be used to determine a confidence interval for the parameter value of interest. A second approach, assuming covariance stationarity, is to estimate the autocovariance function which may then be used to estimate the variance of a sample mean. A related method is to study the spectrum of the process. A fourth method is to identify *regeneration points* in the output realization. These regeneration points form demarcation points defining epochs, behaviour between epochs being statistically independent.

1.3 Simulation in statistics

Simulation is a powerful empirical tool in statistics, to be used when analytical methods fail. For example Thoman *et al.* (1969) use simulation to estimate the distribution of the maximum likelihood estimators of a and b in the two-parameter Weibull density,

$$f_X(x) = \frac{ax^{a-1}e^{-(x/b)^a}}{b^a} \qquad (x \geqslant 0),$$

where $a, b > 0$. Given a random sample $X_1, X_2, \ldots, X_n$, the maximum likelihood estimates are given by the solution to

$$\frac{n}{\hat{a}} + \sum \ln X_i - \sum (X_i/\hat{b})^{\hat{a}} \ln X_i = 0 \qquad (1.2)$$

and

$$n - \sum (X_i/\hat{b})^{\hat{a}} = 0. \qquad (1.3)$$

Now, $Y = (X/b)^a$ is negative exponential, mean 1, so eqns (1.2) and (1.3) may be expressed as

$$\frac{1}{\hat{a}/a} + \frac{\sum \ln Y_i}{n} - \frac{\sum Y_i^{\hat{a}/a} \ln Y_i}{\sum Y_i^{\hat{a}/a}} = 0 \tag{1.4}$$

and

$$\hat{a} \ln(\hat{b}/b) = \ln \left\{ \frac{1}{n} \sum Y_i^{\hat{a}/a} \right\},$$

where $\{Y_i\}$ are n independent negative exponential random variables, mean 1. Since the distribution of $\hat{a}/a$ cannot be deduced analytically from eqn (1.4) the following simulation approach is justified.

A large number m of random samples, each consisting of n unit mean negative exponential variates is drawn by simulation. This gives m values of $\hat{a}/a$ and of $\hat{a} \ln(\hat{b}/b)$. These are now ranked in ascending order so that

$$\frac{\hat{a}_{(1)}}{a} \leqslant \frac{\hat{a}_{(2)}}{a} \ldots \leqslant \frac{\hat{a}_{(m)}}{a},$$

and similarly for $\hat{a} \ln(\hat{b}/b)$.

Considering the former, an empirical or sample distribution function for $\hat{a}/a$ may be obtained using

$$\hat{F}_{\hat{a}/a}(x) = i/m \qquad \left(\frac{\hat{a}_{(i)}}{a} \leqslant x < \frac{\hat{a}_{(i+1)}}{a} \right) \tag{1.5}$$

where $\hat{a}_{(0)} = 0$ and $\hat{a}_{(m+1)} = \infty$.

An important question arising from such a study is the precision attained. Equation (1.5) gives a point estimate for $F_{\hat{a}/a}(x)$. An interval estimate may be obtained by noting that

$$D_m = \max_x |\hat{F}_{\hat{a}/a}(x) - F_{\hat{a}/a}(x)|$$

follows a one-sample Kolmogorov–Smirnov distribution (see also Section 2.4). Thus for $\hat{a}_{(i)}/a \leqslant x \leqslant \hat{a}_{(i+1)}/a$

$$\text{prob}(i/m - D_{m,\alpha} \leqslant F_{\hat{a}/a}(x) \leqslant i/m + D_{m,\alpha}) \geqslant 1 - \alpha$$

where $D_{m,\alpha}$ is the $(1 - \alpha)$ point of the distribution.

The example above *estimated* the distribution of statistics $\hat{a}/a$ and $\hat{a} \ln(\hat{b}/b)$ because the distributions could not be derived analytically. This difficulty often arises in hypothesis testing. However, to provide a good estimate of the distribution of a test statistic would require a very large number of realizations to be generated. Barnard (1963)[†] suggests the following alternative

---

[†] The procedure dates back to around 1950, when it was being taught in University College, London, Statistics courses (I. D. Hill, personal communication, 1987).

procedure. In order to test the significance of some test statistic value t_1, under a null hypothesis H_0, generate via simulation $m-1$ other values (also under H_0) $t_2, \ldots, t_m$. Rank all m values in ascending order such that $t_{(1)} \leqslant t_{(2)} \ldots \leqslant t_{(m)}$. Under H_0 the rank of t_1 will be uniformly distributed in $[1, m]$. Thus for a one-sided test in which large test statistic values are significant, H_0 is rejected at the $100\,\alpha\%$ level if the rank of t_1 exceeds $m(1-\alpha)$. Similar remarks apply to a two-sided test. Hope (1968) showed that the procedure is very effective when uniformly most powerful tests exist, since the loss in power is only slight.

Besag and Diggle (1977) applied the method to testing for spatial randomness. In one example it was required to determine whether 65 Japanese black pine saplings were uniformly distributed in a 5.7-metre square. A X^2 test statistic having value t_1 was constructed by comparing observed and expected frequencies of the 2080 inter-tree distances in $\frac{1}{2}$-metre categories. The dependency between inter-tree distances sharing a common tree means that X^2 cannot be taken to have a chi-squared distribution. In this example m was set to 100, so 99 realizations of the process were generated via simulation. For each of these the X^2 value was calculated. The rank of the test statistic value t_1 was found to be close to the median rank, supporting acceptance of the null hypothesis.

1.4 Simulation languages and new innovations

We have seen that simulation models can be either static or dynamic. In the case of a static system it is usual for the model to be programmed in a *general-purpose language* such as FORTRAN, Pascal, BASIC, ALGOL, or APL. In the case of dynamic systems, however, the experimenter usually has the option of choosing between a general-purpose language and one of the *special-purpose simulation languages*. These include GPSS, GASP, ECSL, SIMULA, and SIMSCRIPT.

The advantage of using a general-purpose language is that the experimenter is likely to be already familiar with it, so no extra time need be invested in learning a new language. Further, such languages are extremely flexible and the complexity of systems which can be modelled is limited only by the ingenuity of the programmer. A disadvantage is that certain routine and recurring tasks involving list processing can be time consuming to program. The special-purpose simulation languages are helpful in this respect, since tasks common to many simulation models are built into the command structure of the language, saving programming time. This feature is usually gained at the expense of some increase in execution time. However, since the cost of execution time relative to programming time is decreasing rapidly with new technology, the net effect may be to favour special-purpose languages. For the frequent designer of simulation models the extra time spent on learning a new language is likely to be justified. There are three areas where simulation languages appear to have failed the serious modeller. Firstly, many such languages do not provide the user with efficient methods for generating random variates from a wide range of distributions. Secondly, few languages use the sophisticated statistical techniques (such as regenerative analysis) necessary for a careful analysis of the output observations. Lastly, it is often difficult for the user to implement experimental design techniques. Even a facility for taking the complement of a random number (the method of *antithetic variates*) may not be available. There seems to be no reason why simulation languages of the future should not incorporate all these features.

A recent innovation is the emergence of *interactive program generators*. These interrogate the user as to the salient features of the system, and from the responses produce a source program in the simulation language. This is then executed in the usual way. The main advantage is that it reduces programming time still further. However, there are limitations as to the complexity of systems which can be modelled. Two such generators are DRAFT (Mathewson, 1977) and CAPS (Clementson, 1982). A review of such generators appears in Mathewson (1975).

One of the advantages of any computer model in business decision making is the ability to answer what-if? questions rapidly. This in turn enables the client to be involved more directly with the use and development of the model. In the early days of simulation, when the client was perhaps remote from the main-frame computer, the use of modems and a telephone line made the computer model 'portable'. The advent of microcomputers has enhanced this portability aspect and has the potential for increasing the involvement of the client. This in turn is likely to lead to more realistic models and hence better decision making. In parallel with the development of microcomputers there has been an increased reliance on the use of graphics to display results, coupled with an interactive component. Hurrion (1976) has described the principles of such *visual interactive simulation*. Further discussion is provided in Hurrion (1978) and Withers and Hurrion (1982).

1.5 Random number and variate generation

The raw material for any stochastic simulation is a sequence of random variates. These in turn are obtained by defined transformations of random numbers. Until the early 1970s there were few methods available for random variate generation. Reliable if slow methods were available for common distributions such as the normal or negative exponential, but other distributions were not well represented. For example, many experimenters used special Erlang approximations to a gamma distribution while no simple yet efficient methods were readily available for the beta distribution. It is a little perplexing to understand why the period 1950–70 was such an inactive one. The probabilistic ideas underlying the algorithms to be described in this book were all well known, simulation was a recognized tool in scientific and management science investigations, and computers were beginning to be used. Indeed it might have been expected that the scarce availability of computing power at the beginning of this period would have motivated the *early* development of efficient algorithms.

The present situation is quite different. There is now a selection of methods available for most standard distributions, and alternative methods continue to be proposed. This is a mixed blessing, since the experimenter is now faced with a decision problem. To solve this, the attributes of a good random variate generator must be identified, together with some sort of weighting to be attached to each attribute. Since the weightings are likely to depend upon the environment in which the generator is used, it will be appreciated that the decision problem is a complex one, which is unlikely to be soluble on a quantitative basis, but may lend itself to qualitative analysis. We consider just five attributes: statistical reliability, set-up time, marginal generation time, program portability/length, and memory requirements.

1.5.1 Statistical reliability

An algorithm is *exact* (rather than approximate), if from a mathematical viewpoint, the variates have precisely the required distribution. Thus, doubling a random number is an exact method of generating variates $\sim U[0, 2)$. Conversely the all too frequent (but not recommended) practice of summing 12 random numbers gives by the central limit theorem a random variable which is only approximately normally distributed. In deciding whether an algorithm is exact or approximate it will be assumed we have a good source of random numbers.

We will deal mainly with exact algorithms. In many simulation experiments uncertainties about how well the model reflects the real world system would justify the use of an approximate algorithm. However, algorithms are frequently the basis of programs inserted into scientific subroutine libraries. In these cases the ultimate use of the routine is often unknown, and it must be assumed, therefore, that good quality variates are required.

Whenever an exact mathematical algorithm is translated into a subroutine, numerical errors will creep in, causing the output from the generator to show statistical properties somewhat different from those intended. The generator is statistically reliable if these differences are sufficiently small. The source of these errors include faults in the random number generator, limited word size in the computer, poor numerical practice by the programmer, and of course logical errors in the program. None of these invalidate the exactness of the original algorithm. Whether these discrepancies are serious or not may be determined by statistical quality control on the output of the generator. Given a sample of output, even quite significant discrepancies may go undetected. Given enough patience and a large enough sample, discrepancies are likely to be identified, even for a relatively good generator, since the null hypothesis can never be exactly true. How large should the sample size be? One approach is to ask how many random variates are to be sampled within a particular simulation. Providing discrepancies are undetectable for such a sample size, the generator would be deemed adequate for that application. Unfortunately this approach means that the sample size could be several million. It does not seem to have been the practice to submit generators to such stringent tests. Perhaps the reason for this is that even a good generator has a high chance of failing such a test. Happily nature usually comes to our assistance, since measures of performance in many systems are often quite robust with respect to small deviations in the distributional properties of the input parameters. To establish statistical reliability of a generator it is necessary to demonstrate that the observations are independently distributed, and are drawn from the required distribution. If the random number (uniform) generator is a good one, then the independence between random variates is implied. Methods for testing random number streams are described in Chapter 2. This leaves a *test of fit* problem for the random variates. Methods based on chi-squared and Kolmogorov–Smirnov distributions are frequently used. Such methods applied to uniform variates are described in Section 2.4 and are easily modified to cope with other distributions. Stephens (1974) gives a comparison of methods. Also of interest are tests available for particular distributions. In particular the dispersion test for the Poisson distribution (Cochran, 1954) requires no grouping of observations in the tail of the distribution.

1.5.2 Set-up time

The set-up time for a generator is the processing time required before the first variate can be generated. It arises when certain preliminary values have to be calculated. For example, it may be necessary to produce a table of constants giving the cumulative distribution function (c.d.f.) at selected values. In simulations where the parameter values of the distribution do not change between successive calls of the generator, a fairly high set-up time can be tolerated when a large number of variates are to be generated. Conversely applications where the parameter values change between calls require a generator with small set-up time.

1.5.3 Marginal generation time

This is the mean processing time required to generate one variate, once set-up calculations have been performed. In the past, reports of experimental results concerning the behaviour of generators have tended to concentrate on this aspect, to the exclusion of set-up time, although the relative importance of the two attributes depends entirely on the application. The marginal generation time will depend upon the mean frequency with which certain operations are performed within the algorithm and the time required to perform each type of operation. Basic arithmetic operations such as comparison, addition, subtraction, and multiplication are relatively fast, while the evaluation of logarithms and trigonometric or exponential functions tends to take much longer under current technology.

The mean frequency with which each type of operation is carried out is likely to be machine independent, while the processing time is dependent upon the machine, language, and compiler or interpreter used. For this reason a performance ranking of algorithms on a marginal time basis is likely to be dependent upon the particular computing environment. It is useful, therefore, to obtain some machine-independent measures of performance. One approach is to obtain theoretical expressions for the expected frequency of each type of operation. For example, it would be useful to know how many random numbers and logarithmic evaluations a particular generator needs on average to generate one variate. Such measures are necessarily imperfect when related to actual timings on a particular machine. However, they are useful in the following respects. They can show (theoretically) how the relative performance of an algorithm will change as the parameter values of the distribution change. In this way an unpromising algorithm can be excluded at an early stage in its development, without wasting valuable programming time. The measures are also useful in predicting how timings will change if some feature of the computing environment changes. For example, if a faster random number generator is used, a knowledge of the mean frequency of use of random numbers will enable the change in marginal generation time to be predicted. Theoretical measures also provide a basis for understanding and explaining the empirical results of timing experiments. Finally they provide confirmatory evidence for the logical validity of a program, in that observed frequency of random number use could, for example, be compared with the theoretical result. Examples of complexity analysis of algorithms are to be found in Knuth (1972) and Payne (1977).

Marginal generation time will also depend upon the precision with which the routine is implemented. A program executed in double precision is slower than one executed in single precision. The ranking of generators on a time basis may well change substantially as computing technology changes. In the future one may expect functions such as logarithm and sine, and perhaps even random numbers themselves, to be generated directly in the computer hardware. As parallel processors become more prevalent, certain algorithms will lend themselves to parallel (rather than sequential) execution. Finally, although it is advantageous to reduce marginal generation time, it is common experience that the generation time frequently accounts for only a small proportion of the total processing time in a given simulation application.

1.5.4 Portability/program length

Whether a generator is 'portable' or not depends on whether it can be easily transferred from one machine to another with a minimal amount of change. For example, an algorithm working on bit-manipulation is likely to need extensive modification if the word size of the new computer is different from the previous one. While functions such as logarithm, sine, and exponential are widely available, access to numerical integration routines may be more limited. Thus a generator requiring numerical integrations is likely to be more portable if these computations are embedded within the generator itself. Even in this case the precision of the numerical integrations may change slightly from one machine to another.

A short program consisting of a few lines of simple code is likely to enhance the portability of a generator. It can probably be implemented rapidly and is less prone to transcription errors. The shortness of the program will often reflect a simple underlying theory which encourages user confidence.

1.5.5 Memory requirements

The storage requirements for a generator will depend upon the program length, the memory occupied by constants, variables, arrays, and whether single or double precision is used. The cost of memory is now relatively cheap, so high storage requirements are not particularly disadvantageous from a financial viewpoint. However, there is often an association between memory requirements and the portability of the generator. Some generators have high storage requirements because a large number of constants have to be input into the routine and stored. Others have lengthy coding which again would detract from the portability.

1.6 Further reading

It has not been the intention in this chapter to show readers how to write simulation programs. Such information can be found in Fishman (1978), Mitrani (1982), Law and Kelton (1982), Pidd (1984), Ellison and Tunnicliffe Wilson (1984), and for historical interest Tocher (1963). In the main these books emphasize discrete event simulations in an operational research context. For a complementary view Morgan (1984) looks at statistical applications. Applications in economics are described in Naylor (1971).

There are several books showing how to program in a special purpose simulation language. These include Gordon (1975) and O'Donovan (1979) for GPSS, Pritsker (1974) for GASP, Birtwistle *et al.* (1973) and Birtwistle (1979) for SIMULA, and Kiviat *et al.* (1969) for SIMSCRIPT.

The Monte Carlo method is generally accepted to date from the early 1940s, from work on the atomic bomb. Early users of the technique included Fermi, Ulam, Metropolis, and von Neumann. For an insight into early applications see Meyer (1956) and Hammersley and Handscombe (1964). More recently Parker (1972) provides a comprehensive review of applications in neutron physics. Earlier instances of Monte Carlo applications under a different guise are Buffon's needle problem in the nineteenth century (Hammersley and Handscombe 1964, p. 6) while W. S. Gossett ('Student' 1908) conducted sampling experiments as part of his work on the t-distribution.

Experimental design techniques in simulation are described in Kleijnen (1974, 1975); these include variance reduction methods. Other useful texts for the latter are Hammersley and Handscombe (1964) and Rubinstein (1981), Fishman (1973, 1978) gives a useful account of the analysis of output data, while Crane and Lemoine (1977) is informative on the regenerative method. Bibliographies on random number and variate generation have been compiled by Sowey (1972, 1978) and Sahai (1979).

There is a vast literature on applications of simulation. Holst (1979) gives a bibliography with emphasis on the applications. Articles on applications and methods appear with varying degrees of frequency in the following journals: *Journal of the Operational Research Society, Management Science, Communications in Statistics (Part B), Operations Research, European Journal of Operational Research, Technometrics, Biometrica, Biometrics, Journal of the Association of Computing Machinery, Mathematics of Computing, Computing, Journal of the American Statistical Association, American Statistician, SIAM Journal in Scientific and Statistical Computing, and Computers and Mathematics with Applications.*

2
Random number sequences

2.1 Introduction

Most simulations involve some random or chance elements, so there is a need for a sequence of *random* numbers. An appropriate definition for such a sequence is not at all obvious. A thorough discussion leading to definitions which seem in accord with our intuitive ideas as to the meaning of random is given in Knuth (1981). For our purposes it will suffice to highlight the main features we expect a random number sequence to exhibit. A distinction needs to be made between infinite and finite sequences. We will consider infinite sequences only and the reader is referred to Knuth (1981) for a treatment of finite sequences. Given an infinite sequence $R_0, R_1, \ldots, R_n, \ldots$ most people's notion of random would include at the very least that the numbers be uniformly distributed in the interval $[0, 1)$.[†] Such a sequence is said to be *1-distributed*. It is easy to imagine sequences which are 1-distributed, but are in no sense random, because there is some form of dependence. For example, a high number may consistently be followed by a low number and vice versa. To prevent such sequences being called random, we might further insist that the sequence of 2-tuples $\{(R_n, R_{n+1})\}$ be uniformly distributed in the unit square, that is that the sequence be 2-distributed. In general, a sequence is *k-distributed* if the *k*-tuples $\{(R_n, R_{n+1}, \ldots, R_{n+k-1})\}$ are uniformly distributed in the *k*-dimensional unit hypercube $[0, 1)^k$. Clearly if a sequence is *k*-distributed then it is $(k-1)$-distributed.

To discover whether *k*-distributivity is in itself a strong enough requirement for randomness, consider Fig. 2.1. Plot (a) shows the distribution of 500 2-tuples $\{(R_n, R_{n+1})\}$ arising from a certain generator. The generator is of the Fibonacci type (see Section 2.3.1) with modulus 125 and starting values 0 and 1. Although there are four isolated regions of higher than average density, there is evidence of a very crude uniformity over the unit square, showing that the sequence possesses a degree of 2-distributivity. Plots (b)–(e) each show the distribution of sub-sequences formed by taking every fifth 2-tuple. Since the *period* of the generator is 500, there are five such sub-sequences, each yielding

---

[†] Arguably the interval (0, 1) would be preferable, to avoid problems associated with dividing by or taking the logarithm of zero. For a further discussion, see Section 2.4.9.

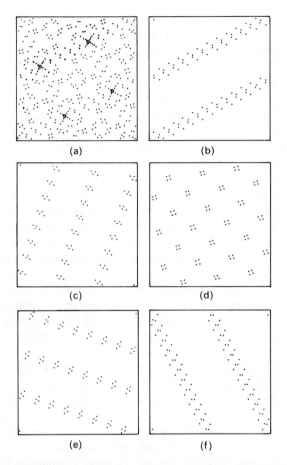

Fig. 2.1. The distribution of (a) $\{(R_n, R_{n+1})\}$, (b) $\{(R_{5n}, R_{5n+1})\}$, (c) $\{(R_{5n+1}, R_{5n+2})\}$, (d) $\{(R_{5n+2}, R_{5n+3})\}$, (e) $\{(R_{5n+3}, R_{5n+4})\}$, (f) $\{(R_{5n+4}, R_{5n+5})\}$ for a Fibonacci generator modulus 125

100 2-tuples. Plot (d) possesses a degree of uniformity, (c) and (e) are very marginal in this respect, while (b) and (f) are obviously non-uniform. A sequence is said to be (h, k)-*distributed* if every sequence obtained by taking every hth k-tuple is uniformly distributed over $[0, 1)^k$. Thus the visual evidence for this sequence suggests that it is not $(5, 2)$-distributed but is (to a low order of precision) 2-distributed. Of course *absolute* k- or (h, k)-distributivity is not attainable in practical generators. The Fibonacci sequence yielded only 500 attainable values for (R_n, R_{n+1}), so it is not possible to obtain 2-distributivity to an order of accuracy better than 1 part in $\sqrt{500}$, $1/\sqrt{500}$ representing the average distance between points in (a). Similarly

(5, 2)-distributivity could not be obtained to an accuracy of better than 1 part in $\sqrt{100}$ or 1 part in 10. Note how these plots can provide a useful sufficient condition for rejecting a number sequence from a short period generator. All the plots (a)–(f) display undesirable features and it would be most inadvisable to use any of them for a practical generator.

Just as k-distributivity implies $(k-1)$-distributivity, so a (h, k)-distributed sequence implies $(h, k-1)$-distributivity. Another useful property is that it is also (μ, k)-distributed whenever μ divides h, which shows that the sequence is also $(1, k)$- or k-distributed. Ideally a random sequence should be k-distributed for all k, that is, it should be ∞-*distributed*. It transpires that an ∞-distributed sequence is also (h, k)-distributed for all h and k (Knuth, 1981, p. 149), so it might appear that ∞-distributivity is sufficient for randomness. Unfortunately it is not (at least according to our intuitive notions) since it is still possible for certain subsequences to display non-random behaviour. The ∞-distributivity has therefore to be strengthened by additional requirements, relating to the 1-distributivity of certain infinite subsequences (Knuth, 1981, pp. 152–6).

How are random number sequences generated? Early attempts included the use of physical devices such as a roulette wheel or the random noise in an electronic circuit. More recently physical devices have utilized the emission of γ-rays from radioactive nuclides (Inoue *et al.*, 1983). Paradoxically, most simulation experiments require the sequence to be *reproducible*, in order to assist in the debugging of computer programs, and to perform controlled experiments when the response of a system under different policy options is to be investigated. Reproducible sequences are known as *pseudo*-random number streams, and these are the ones that are used almost exclusively in simulation today. One method of producing them is to use a physical device and then store the numbers; a notable early example of this is 'A Million Random Digits with 100,000 Normal Deviates', published by the Rand Corporation in 1955. For a computer simulation, the numbers generated from a physical device can be stored on tape or disk. This ensures reproducibility, but is usually cumbersome, causing increased memory requirements and/or transfer time from the peripherals.

For this reason, pseudo-random number streams (or simply random number streams as we shall call them from now on) are usually generated at run-time, using a deterministic recurrence equation. The recurrence equation ensures that the stream is reproducible. The deterministic nature of generation means that the designer has to be painstaking in the choice of the precise relationship of the recursion, otherwise unwanted dependencies will appear. Ideally a random number generator would be designed for each simulation application, and would be tested extensively to ensure that the inevitable dependencies that do exist do not have an undesirable effect on the model of the system being simulated. Few experimenters could justify the effort required to carry out these tests. In practice, therefore, random number generators are tested for particular deviations from randomness, regardless of the application for which they may be used. Given the diversity of such applications, the stringency of tests has to be much greater. It does not seem unreasonable for the simulation experimenter to expect that the generator available on a particular machine is a good one, and for the supplier to be aware of this responsibility. There have been infamous cases of a random number generator being used extensively despite discovery of distinctly non-random features. Perhaps the best known of these is RANDU, the IBM-based generator (see, for example, Ripley, 1983a).

RANDU is one of the poorer examples of the *linear congruential* method of obtaining random numbers. This class of methods has been used for over two decades, and the vast majority of computer installations have their random number generators based on it. The next section discusses the properties of this important method.

2.2 Linear congruential generators

Linear congruential generators deliver a sequence of non-negative integers $\{X_i\}$ where

$$X_{i+1} = (aX_i + c) \bmod m \qquad (i = 0, 1, \ldots). \tag{2.1}$$

m is a specified positive integer known as the *modulus*; X_0, a, and c are specified non-negative integers known as the *seed*, *multiplier*, and *increment*, respectively. They take values in the range $[0, m-1]$. The modulo m process returns the remainder after dividing by m, and so $X_i \in [0, m-1]$. The 'random number' is delivered as $R_i = X_i/m$, and so $R_i \in [0, 1)$.

We define the *period* of the generator to be the smallest positive integer λ for which $X_\lambda = X_0$. The sequence repeats itself after λ numbers. It is in our interests to make the period as large as possible. This will enhance the number of attainable values for k-tuples, and hence improve the chance of obtaining higher-order distributivity. Clearly the period can be no greater than m, so, subject to other competing requirements (such as computational efficiency), m is often chosen to be equal or nearly equal to the largest representable integer in the computer.

A *full period* generator is one in which the period is m. Hull and Dobell (1962) showed that a full period is obtained if and only if:

(1) c is relatively prime to m;
(2) $(a-1)$ is a multiple of q, for each prime factor q of m;
(3) $(a-1)$ is a multiple of 4, if m is.

Linear congruential generators can be classified into *mixed* $(c > 0)$ and *multiplicitive* $(c = 0)$ types.

2.2.1 Mixed generators

The conditions above are certainly satisfied if m is a prime number and $a = 1$. In this case $X_{i+1} = X_i + c$ or $X_{i+1} = X_i + c - m$, so the 2-tuples $\{(X_i, X_{i+1})\}$ lie on one of only two lines. Thus, despite its computational advantages (it has unit multiplier), such a sequence has very poor 2-distributivity, and cannot seriously be considered.

For binary computers, the case $m = 2^b$, where b is some positive integer, is of considerable interest. This is because the modulo m process can be performed very efficiently merely by retaining the lower b bits in the binary representation of $aX_i + c$. Such a generator $(b \geqslant 2)$ will have full period if $(a - 1)$ is a multiple of 4 (satisfying conditions (2) and (3) above) and c is odd-valued (satisfying condition (1)). Since we wish to maximize the period m, b is usually set to the number of bits used to represent integers on the computer. For example, on an IBM 370, which is a 32-bit computer, b may be set to 31 (one bit reserved for sign designation) or to 32 (using 2's complement arithmetic). In the latter case the period will be $2^{32} = 4.29 \times 10^9$. For a 16-bit personal computer, where we set $m = 2^{15}$, the period reduces to 32 768. Thus it would be impossible to achieve k-distributivity to better than 1 part in $\sqrt[k]{32\,768}$, illustrating the advantages of long word-length computers. A disadvantage of choosing a power of 2 modulus, is that the lower $d \, (< b)$ bits have period no greater than 2^d. For example, the last 8 bits have a period no greater than 256. Providing the experimenter is not extracting these low-order bits for use as random numbers this should not be too serious. Note that reproducibility is guaranteed by specifying the same seed value on separate occasions. Where the experimenter wishes to obtain different parts of the sequence for different experiments, the seed value can be chosen as the last integer value X_i used on the previous run, or it can be selected randomly by the internal clock of the computer system. In the latter case there is a small risk of overlap between the separate streams used.

2.2.2 Multiplicative generators

In this case

$$X_{i+1} = (aX_i) \bmod m. \tag{2.2}$$

Since $c = 0$, condition (1) of the full period requirement is never satisfied. A full period is therefore not possible. This is also apparent from the requirement that $X_i \neq 0$, lest the sequence become degenerate. Given m, it is natural to choose a multiplier a which maximizes the period. For given m, we call this a *maximum period* generator. If X_i and m have a common divisor d, application of eqn (2.2) shows that X_{i+1} will also have a divisor d. All numbers in the sequence will therefore be multiples of d. This severely reduces the period of the generator d. It is natural, therefore, to insist that X_i be relatively prime to m for all i. It follows that the period can be no greater than the number of integers in $[0, m-1]$ which are relatively prime to m. One obvious choice for m is therefore a prime number.

If m is prime, then a maximum period generator is one in which a is a *primitive element modulo m* (Knuth, 1981, pp. 19–21). In this case the period $\lambda = m-1$. A primitive element modulo m is one such that $a \bmod m \neq 0$ and $\{a^{(m-1)/q}\} \bmod m \neq 1$ for each prime factor q of $m-1$. Given preceding comments, m is often set to the largest prime number less than 2^b, where b is the number of usable bits. This *prime modulus generator* has period $m-1$, and is often also known as an *almost full period generator*. An advantage over the full period multiplicative generator is that the low-order bits tend to be more random. A disadvantage is that finding primitive elements modulo m can be difficult, although if a is such an element, then so is a^k. The number of possible choices for a is the number of integers in $[0, m-1]$ which are relatively prime to $m-1$. An apparent disadvantage is that the modulo m process cannot be achieved as efficiently as with $m = 2^b$. Fortunately steps can be taken (Fishman, 1978, p. 357) to emulate the bit shifting process by representing m as $2^b - \gamma$. The method is particularly efficient when $\gamma = 1$. Primes of the form $2^b - 1$ are *Mersenne primes*, the low-order Mersenne exponents being $b = 2$, 3, 5, 7, 13, 17, 19, 31, 61, 89, 107, ...

Another choice of modulus is $m = 2^b$. In this case the maximum period generator is one in which $a = 3 \bmod 8$ or $5 \bmod 8$ and X_0 is odd (Knuth, 1981, pp. 19–21). The period $\lambda = m/4$, which is a crucial disadvantage compared with the full period multiplicative and prime modulus generators. All numbers in the sequence will be odd-valued, which accounts for the absence of 50 per cent of the numbers in $[0, m-1]$. For fixed a and m, two non-communicating cycles of $m/4$ odd numbers are obtainable from the generator. These are a permutation of $\{1, 3, 9, \ldots\}$ and $\{5, 7, 13, \ldots\}$ for $a = 3 \bmod 8$ or $\{1, 5, 9, \ldots\}$ and $\{3, 7, 11, \ldots\}$ for $a = 5 \bmod 8$. It transpires (Ripley, 1983b) that if $a = 5 \bmod 8$, the sequence can be obtained directly from a full period mixed generator with identical multiplier, increment $(X_0 \bmod 4)(a-1)/4$, and modulus $m/4$. To demonstate this, we have

$$X_{i+1} = aX_i \bmod 2^b.$$

Let

$$d = X_0 \bmod 4.$$

Then

$$X_{i+1} - d = \{a(X_i - d) + (a-1)d\} \bmod 2^b. \tag{2.3}$$

If $a = 5 \bmod 8$, then $(a-1)d$ is a multiple of 4. Suppose now that $(X_i - d)$ is a multiple of 4. Then induction on i in eqn (2.3) gives $(X_i - d) \bmod 4 = 0$ for all i. Thus $(X_i - d)/4 = Y_i$ for some integer Y_i. Substitution into eqn (2.3) yields

$$Y_{i+1} = \left\{aY_i + \frac{(a-1)d}{4}\right\} \bmod (m/4)$$

where
$$X_i = 4Y_i + X_0 \bmod 4.$$

Dividing by m gives a sequence of random numbers $\{R_i\}$ where
$$R_i = R_i' + (X_0 \bmod 4)/m$$

and $\{R_i'\}$ is the sequence produced from the associated full period generator (modulus $m/4$). In view of the link between modulo 2^b multiplicative generators with $a = 5 \bmod 8$ and full period generators, we will in the main consider the multiplicative type with the restriction $a = 5 \bmod 8$.

2.2.3 Theoretical tests of randomness

Most linear congruential generators available today fall into one of three categories described above. We summarize these as

type A   full period mixed, $m = 2^b$, $a = 1 \bmod 4$, c odd, $\lambda = m$;

type B   maximum period prime modulus, $m =$ prime, $a =$ primitive element modulo m, $c = 0$, $\lambda = m - 1$;

type C   maximum period multiplicative, $m = 2^b$, $a = 5 \bmod 8$, $c = 0$, $\lambda = m/4$.

The aim of *theoretical tests* is to determine, solely from the structure of the generator, whether particular values of a and c lead to non-random behaviour. Theoretical tests are complementay to *empirical tests* which examine a sample of the output from a generator and subject it to a number of statistical tests. Most of the theoretical results relate to generators of type A. Fortunately C is subsumed by A, while the absence of $X_i = 0$ in B requires only minor changes in the analysis. The theoretical properties relate to the detection of *global* non-randomness, i.e. behaviour over the entire cycle $(X_0, \ldots, X_{\lambda-1})$. This says little about local non-randomness, the detection of which must be left in the main to the empirical testing procedures (Section 2.4).

2.2.3.1 *Mean, variance, and first-order serial correlation*

For a random number sequence $\{R_i\}$ we hope that $E(R_i) = 0.5$, $V(R_i) = 1/12$, and $\rho(R_i, R_{i+1}) = 0$. Since $R_i = X_i/m$, these are equivalent to $E(X_i) = m/2$, $V(X_i) = m^2/12$, and $\rho(X_i, X_{i+1}) = 0$. For a type A, B, or C generator

$$\mu = E(X_i) = \frac{1}{\lambda} \sum_{i=0}^{\lambda-1} X_i$$

and

$$\sigma^2 = V(X_i) = \frac{1}{\lambda} \sum_{i=0}^{\lambda-1} X_i^2 - \mu^2.$$

For a type A generator $\lambda = m$ and $\{X_i\}$ is a permutation of $(0, 1, \ldots, m-1)$ giving

$$\mu = \frac{1}{m} \sum_{n=0}^{m-1} n = \frac{m-1}{2}$$

and

$$\sigma^2 = \frac{1}{m} \sum_{n=0}^{m-1} n^2 - \left(\frac{m-1}{2}\right)^2$$

$$= \frac{m^2 - 1}{12}.$$

For large m, these are perfectly acceptable compared with $m/2$ and $m^2/12$. For a type B generator equivalent analysis yields $\mu = m/2$ and $\sigma^2 = (m^2 - 2m)/12$.

Proceeding in a similar manner, for type A only, the first-order serial correlation between X_i and X_{i+1} is

$$\rho = \frac{\dfrac{1}{m} \sum_{i=0}^{m-1} X_i X_{i+1} - \mu^2}{\sigma^2}$$

$$= \frac{\sum_{n=0}^{m-1} 12nS(n) - 3m(m-1)^2}{m(m^2-1)}$$

where $S(n) = (an + c) \bmod m$. Coveyou (1960) found the bounds

$$\left| \rho - \frac{1}{a}\left(1 - \frac{6c}{m} + 6\left(\frac{c}{m}\right)^2\right) \right| < \frac{a+6}{m}. \tag{2.4}$$

Unfortunately result (2.4) has been used extensively to select 'good' values of a and c for specified m. In fact, it is of little use, since the bound is too large to investigate values of a and c giving really low first-order serial correlations. It turns out that serial correlations of the order $\pm 100/m$ are easily obtained, yet this is not deducible from eqn (2.4). An improved approximation was given by Greenberger (1961) but suffered from the same problem. Dieter (1971) and Knuth (1977) obtained far more precise bounds. These are based on Dedekind sums which depend upon a and m, but not on c. The importance of this work is that if a multiplier a, which is relatively prime to m, is chosen at random, the serial correlation is likely to be extremely small. Of course this does not mean that a randomly chosen multiplier will give a good generator, rather that low first-order serial correlation is hardly any more difficult to obtain than the correct mean and variance. A final remark is that the multiplier a should be neither too small nor too large. The equation

$$X_{i+1} - aX_i = X_{i+1} + (m-a)X_i = c \bmod m$$

shows that the set of all 2-tuples can be covered by a set of few parallel lines if a or $m-a$ is small. This gives poor 2-distributivity. Knuth (1981, p. 170) suggests that a be chosen between $0.01m$ and $0.99m$.

2.2.3.2 *Potency*

The potency of a linear congruential generator is defined to be the smallest integer s for which $(a-1)^s \bmod m = 0$. This measure was introduced by Knuth (1981, pp. 22–4). For a full period generator the potency must exist since $(a-1) \bmod q = 0$ for each prime factor q of m. For any linear congruential generator

$$X_{i+1} = (aX_i + c) \bmod m \qquad (2.5)$$

we have

$$X_{i+1} - X_i = \{a(X_i - X_{i-1})\} \bmod m.$$

Repeated application of eqn (2.5) gives

$$X_{i+1} - X_i = \{a^i(X_1 - X_0)\} \bmod m.$$

Consider the case of a full period generator. Without loss of generality we may take $X_0 = 0$ (it must appear somewhere in the sequence) giving $X_1 = c$ and hence

$$X_{i+1} - X_i = \{a^i c\} \bmod m.$$

Let $d = a - 1$. Then

$$X_{i+1} - X_i = \{(d+1)^i c\} \bmod m$$

$$= \left\{ \sum_{j=0}^{i} \binom{i}{j} d^j c \right\} \bmod m. \qquad (2.6)$$

For $i \geqslant s$, this becomes

$$X_{i+1} - X_i = \left\{ \sum_{j=0}^{s-1} \binom{i}{j} d^j c \right\} \bmod m, \qquad (2.7)$$

since $d^j \bmod m = 0$ when $j \geqslant s$. If the potency s is 1, eqns (2.6) and (2.7) yield

$$X_{i+1} - X_i = c \bmod m,$$

implying that all 2-tuples (X_i, X_{i+1}) lie either on the line $X_{i+1} - X_i = c$ or on $X_{i+1} - X_i = c - m$. Such a sequence therefore has extremely poor 2-distributivity. If the potency is 2, eqns (2.6) and (2.7) now give

$$X_{i+1} - X_i = \{c(1 + id)\} \bmod m.$$

Replacing i by $(i-1)$ and differencing yields

$$X_{i+1} - 2X_i + X_{i-1} = (cd) \bmod m,$$

indicating that all 3-tuples (X_{i-1}, X_i, X_{i+1}) lie on one of four planes. Thus such a sequence possesses poor 3-distributivity. In a similar vein we may show that generators with potency 3 and 4 satisfy

$$X_{i+1} - 3X_i + 3X_{i-1} - X_{i-2} = (cd^2) \bmod m$$

and

$$X_{i+1} - 4X_i + 6X_{i-1} - 4X_{i-2} + X_{i-3} = (cd^3) \bmod m$$

respectively. In the first case the 4-tuples lie on no more than eight planes and in the second case the 5-tuples lie on no more than 16 planes. These examples illustrate that low potency (in the case of full period generators) implies non-random behaviour.

Knuth suggests that the minimum potency for such generators should be 5. We may find values of a which maximize the potency for type A generators. Since $a = 1 + 4v$ for some integer v, we must find the smallest s, such that $(4v)^s \bmod 2^b = 0$. Taking $b \geqslant 2$, a sufficient condition for the potency to be maximized is that v be odd-valued in which case the potency is $b/2$ (b even) or $(b+1)/2$ (b odd). Thus for maximum potency $a = 5 \bmod 8$, and we may therefore replace the requirement $a = 1 \bmod 4$ by this (stricter) condition to improve our chances of obtaining a good sequence. Multipliers of the form $a = 2^k + 1$ (type A generators) are usually unsatisfactory, since the potency is the smallest s for which $ks \geqslant b$. For example, if $b = 32$, then a potency of at least 5 is achievable only if $k \leqslant \langle 32/5 \rangle + 1 = 7$ or $a = 129, 65, 33, 17, 9, 5$. However, as we noted previously, generators with small a/m tend to exhibit poor 2-distributivity. Thus for type A generators, multipliers $a = 2^k + 1$ are effectively ruled out.

Potency is only one way of predicting that low-order k-tuples lie on few planes—a sign of poor k-distributivity. We can often spot similar behaviour using *ad hoc* methods. Consider, for example, a multiplicative generator with $m = 2^b$, $a = 2^n + 3$ where $2n \geqslant b$. Then

$$X_{i+1} = (2^n + 3)X_i \bmod 2^b$$
$$= (2^{2n} + 9 + 6.2^n)X_{i-1} \bmod 2^b$$
$$= (9 + 6.2^n)X_{i-1} \bmod 2^b$$
$$= (6X_i + 9X_{i-1} - 18X_{i-1}) \bmod 2^b.$$

Thus all 3-tuples lie on one of 15 planes

$$X_{i+1} - 6X_i + 9X_{i-1} = 0 \bmod 2^b.$$

Examples of implementations possessing these very poor properties are RANDU ($a = 2^{16} + 3$, $m = 2^{31}$) and Greenberger (1965) ($a = 2^{18} + 3$, $m = 2^{35}$).

Such quick methods are useful for spotting poor generators, but not sufficient for identifying good ones. To be more effective in the latter we need to take a detailed look at the *lattice* structure of linear congruential generators.

2.2.3.3 *Lattice behaviour*

Consider a full period generator giving a sequence $(X_0, X_1, \ldots, X_{m-1})$ and associated random numbers $(R_0, R_1, \ldots, R_{m-1})$. From such a sequence we may consider the distribution of the k-tuples $\{(R_i, \ldots, R_{i+k-1})\}$ through $[0, 1)^k$.

Figure 2.2 shows the distribution of the 256 2-tuples of the generators $R_i = X_i/256$ where

$$X_{i+1} = (5X_i + 3) \bmod 256$$

and

$$X_{i+1} = (13X_i + 3) \bmod 256$$

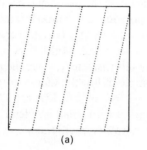 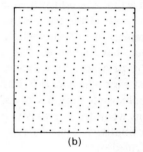

(a)   (b)

Fig. 2.2. Distribution of $\{(R_i, R_{i+1})\}$ for (a) $X_{i+1} = (5X_i + 3) \bmod 256$, (b) $X_{i+1} = (13X_i + 3) \bmod 256$

respectively through the unit square $[0, 1)$. In each case the points may be covered by a family of parallel lines having constant separation. Several such families exist, but in Fig. 2.2(a) the minimum number of lines required to cover all points is five, while in Fig. 2.2(b) it is 13. So in general it appears that the larger is the minimum numer of hyperplanes required to cover all points, the better will be the distribution of points. The distance between neighbouring planes ($1/v$) is a constant, and this distance is necessarily larger in Fig. 2.2(a), because it has fewer planes covering the points. In both cases the points form a regular *lattice*. From the lattice, parallelograms of various sizes may be constructed. If we take the smallest such parallelogram ABCD (Fig. 2.3), it will be noticed that the ratio of side lengths $l_2:l_1$ (larger to smaller) is approximately 10 in Fig. 2.2(a) and about 1.5 in Fig. 2.2(b). (The exact values are 9.84 and 1.56, respectively, obtained from algorithm (2.1).) Thus the more uniform the distribution of 2-tuples, the closer the ratio of side lengths is to 1. Note that there are 256 parallelograms and so the volume of each one is 1/256. It follows that l_1 (the smaller side length) and $1/v$ (the separation between hyperplanes) satisfy the relationship $l_1/v = 1/256$ (Fig. 2.3).

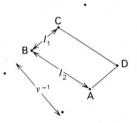

Fig. 2.3. Volume of lattice cell, related to side length (smallest) and separation

Clearly $l_1 l_2 \geqslant 1/256$ and $l_2 v \geqslant 1$, with equality if the parallelogram is rectangular. The latter shows that if the parallelogram or basic lattice cells do not deviate too much from rectangles, then l_2 and v are approximately *reciprocal* measures of uniformity.

A great deal of work has been performed on the lattice structure of type A, B, and C generators. The motivation for understanding lattice structure is the search for 'good' multipliers a (it transpires that the value of c is largely immaterial). The discovery of the lattice is attributed to Marsaglia (1968) in an aptly named paper 'Random Numbers fall mainly in the planes'. Early tests of generators are due to Beyer *et al.* (1971), Marsaglia (1972) (the *lattice test*), and Coveyou and Macpherson (1967) (the *spectral test*). Golder (1976a, b) implemented the latter, although the conditions of applicability were incorrectly stated (see, for example, Hoaglin and King, 1978). Theoretical work on the structure of lattices is given in Dieter (1975), Knuth (1981), and Ripley (1983b), while an up-to-date implementation of the spectral test is given in Hopkins (1983). A brief description of the main results now follows.

The set of k-tuples produced by a full period (type A) generator $X_{i+1} = (aX_i + c) \bmod m$ is

$$\{R_k\} = \{(R_i, R_{i+1}, \ldots, R_{i+k-1})\}$$
$$= \left\{ \left(\frac{0}{m}, \frac{c}{m}, \frac{ac+c}{m}, \ldots \right) \right.$$
$$\left(\frac{1}{m}, \frac{a+c}{m}, \frac{a^2+ac+c}{m}, \ldots \right)$$
$$\vdots \qquad\qquad \vdots$$
$$\left. \left(\frac{m-1}{m}, \frac{a(m-1)+c}{m}, \frac{a^2(m-1)+ac+c}{m}, \ldots \right) \right\} \bmod 1 \qquad (2.8)$$

where we have used $X_i = 0, 1, \ldots, m-1$ in turn to generate the entire set. Defining

$$x = \frac{c}{m}(0, 1, 1+a, \ldots),$$

$$\{R_k\} = \left\{ \left[x + \left(\frac{t_1}{m}, \frac{t_1 a}{m}, \frac{t_1 a^2}{m}, \ldots \right) \right] \bmod 1 \right\}$$

where $0 \leqslant t_1 \leqslant m$ and t_1 is integer-valued.
Thus

$$\{R_k\} = [0, 1)^k \cap \left\{ x + \left(\frac{t_1}{m}, \frac{t_1 a}{m} + t_2, \frac{t_1 a^2}{m} + t_3, \ldots \right) \right\}$$

where $\{t_i\}$ are integer-valued. We may write

$$\{R_k\} = [0, 1)^k \cap \{x + \Lambda\}$$

where Λ is a set of vectors (the lattice)

$$\Lambda = \sum_{i=1}^{k} t_i e_i, \qquad (2.9)$$

and

$$e_1 = (1, a, a^2, \ldots, a^{k-1})/m$$

$$e_2 = (0, 1, 0, \ldots, 0)$$

$$e_3 = (0, 0, 1, \ldots, 0)$$

$$\vdots$$

$$e_k = (0, 0, 0, \ldots, 1).$$

Note that the $\{e_i\}$ form a set of linearly independent basis vectors for the lattice set Λ. The increment c affects the set $\{R_k\}$ only through x. Thus different c serve only to translate the points $\{R_k\}$, but not to change their relative positions. Since the tests relate to the latter, the increment c has no effect on the tests, supporting our earlier claim that the value of c is not crucial to the choice of a good generator.

Equation (2.9) provides a description of the lattice for a type A generator. The results for type C follow by replacing m by $m/4$ and c by $(X_0 \bmod 4)$ $(a-1)/4$. But c has no effect on Λ, so the only change is to replace m by $m/4$ in Λ. For the prime modulus generator, type B, we note that $X_i \neq 0$, so the first member of the set (eqn (2.8)) is absent, giving $\{R_k\} = (0, 1)^k \cap \Lambda$. The lattice set Λ is therefore identical for full period multiplicative and prime modulus generators, having the same multiplier a. Finally for non-overlapping k-tuples the type A generator also produces a lattice with m replaced by $m/\text{g.c.d.}$ (k, m) (Ripley 1983b).

The lattice test aims to identify the smallest parallelepiped in k dimensions (it will have volume m^{-1}) and the associated side lengths $l_1, l_2, \ldots, l_k$, ranked in ascending order. The quantity $r_k = l_k/l_1$ is a measure of how close the parallelepiped is to a k-dimensional hypercube, the latter being associated with the most uniform distribution possible for the k-tuples. A value of r_k close to 1 is therefore aimed for. The very best random number generators have $r_k \leqslant 2$ for $2 \leqslant k \leqslant 5$ (see, for example, Table 3 of Atkinson, 1980).

The spectral test determines a family of hyperplanes covering the lattice with the minimum number of planes and then measures the separation $1/v_k$ between neighbouring planes. For $2 \leqslant k \leqslant 6$ Knuth (1981, p. 101) suggests that $v_k \geqslant 2^{30/k}$ will give an adequate generator. In order to compare spectral values independently of m, Knuth suggests the test quantity

$$\mu_k = \frac{\pi^{k/2} v_k^k}{m\Gamma(1 + k/2)},$$

with $\mu_k \gtrsim 0.1$ for $2 \leqslant k \leqslant 6$ being evidence of an adequate generator and $\mu_k \gtrsim 1$ for $2 \leqslant k \leqslant 6$ being a very good one.

A third measure for the lattice is the length of the longest side l_k (Ripley, 1983b) which ideally should be of the order of $m^{-1/k}$ (reflecting roughly equal side lengths). The relationship between these three measures is clarified by Ripley who demonstrates that

$$1 \leqslant v_k l_k \leqslant m \prod_{i=1}^{k} l_i \leqslant c_k^k \tag{2.10}$$

where $c_k^{2k} = 4/3, 2, 4, 8, 64/3, 64, 256$ for $k = 2, 3, \ldots, 8$, respectively. For small k the outer bounds of (2.10) are not too widely separated, showing that l_k and $1/v_k$ are attempting to measure similar properties.

The main problem is how to determine l_1, l_k, and v_k. Define a *reduced* basis $\{e_i\}$ for the lattice Λ as one in which e_i is the ith shortest vector in Λ, subject to the restriction that $\{e_i\}$ can be extended to a basis for Λ. Then for $k \leqslant 4$ it turns out that $\{|e_i|\}$ are identical to the required side lengths $\{l_i\}$ of the lattice cell or parallelepiped. The aim, therefore, at least for $k \leqslant 4$, is to find a procedure which generates a reduced basis. For $k = 2$, Algorithm 2.1 (see, for example, Marsaglia, 1972; Atkinson, 1980; Ripley, 1983b) guarantees generation of a reduced basis.

Line S1 evaluates the initial basis. For any pair $(i, j) j > i$, lines 4–12 ensure that $|e_i| \leqslant |e_j|$ (by exchange if necessary) and reduce the length of the larger vector by subtracting an integral multiple of the smaller. Note that in line 8, q is evaluated as the integer closest to s. However, if $s = \pm 0.5$ this could cause an infinite loop with e_j taking values $e_j \mp e_i, e_j \pm e_i, e_j \mp e_i, \ldots$. To avoid this, e_j is changed only if its length becomes strictly smaller, that is if $|s| > 0.5$. If for any $(i, j) j > i$, any exchange (line 5) or reduction (line 9) takes place, then f is set to 1, indicating the necessity for a further pass through lines 2–14.

For $3 \leqslant k \leqslant 4$, the algorithm does not guarantee a reduced basis (Atkinson assumes it does). Ripley (1983b) shows how the output from Algorithm 2.1

Algorithm 2.1

Input $[a, m, k]$
Output $[e_1, \ldots, e_k]$

S1 $e_1 := \dfrac{1}{m}(1, a, a^2, \ldots, a^{k-1})$ mod 1.

    $e_2 := (0, 1, 0, \ldots, 0)$

    ⋮

    $e_k := (0, 0, 0, \ldots, 1)$

1    $f := 0$
2   For $i := 1, k-1$
3   For $j := i+1, k$
4   If $|e_i| \leqslant |e_j|$ goto 6.
5   Exchange e_i and e_j, $f := 1$.
6        $s := e_i' \, e_j / |e_i|^2$.
7   If $|s| > 0.5$ then
8          $q := \langle 0.5 + s \rangle$
9          $e_j := e_j - q e_i$
10         $f := 1$
11   If $|e_i| > |e_j|$ goto 5.
12   End If.
13   Next j
14   Next i.
15   If $f = 1$ goto 1.
16   Exit.

can be modified to make it a reduced basis. Fortunately such modifications may not be necessary in practice. Bounds on v_k and l_k can be determined for *any* basis $\{e_i\}$ in which the vectors are ranked in ascending size. Ripley shows that

$$\frac{1}{w}, \; 1 \bigg/ \left\{ m \prod_{1}^{k-1} |e_i| \right\} \leqslant l_k \leqslant |e_k|$$

and

$$\frac{1}{|e_k|} \leqslant v_k \leqslant w \qquad\qquad (2.11)$$

where

$$w = \min_i |e_i^*| \quad \text{and} \quad e_i^* \cdot e_j = \delta_{ij}.$$

The $\{e_i\}$ obtained from Algorithm 2.1 have been reduced substantially, and so the bounds are quite tight. Ripley found that for $k \leqslant 6$, $w|e_k| \leqslant 1.3$ on most occasions. The importance of this is that result (2.11) may be used in conjunction with Algorithm 2.1 to produce a fairly tight lower bound on v_k (it should be large) and a corresponding upper bound on l_k (it should be small). Thus a reasonably precise estimate of the quality of the generator can be obtained without exhaustive calculations. Ripley shows that if

$$m \prod_{i=1}^{k} |e_i| < 2$$

in Algorithm 2.1, a reduced basis has been reached. Of course for $k = 2$, $|e_k|$ and l_k are identical because the $\{e_i\}$ form a reduced basis. The exact value of v_k for general $k \leqslant 9$ may be obtained (with some programming effort) using Knuth's implementation of the spectral test (1981, pp. 94–101). This has been performed by Hopkins (1983).

We conclude this section with a selection of good multipliers. Table 2.1 shows l_k and the ratio $r_k = l_k/l_1$ for four such examples, as computed by Ripley. Also shown is $m^{-1/k}$ ($(m/4)^{-1/k}$ for type C generators), which is a measure of the best achievable longest side length in k dimensions.

Examples of bad generators abound in the literature. For example, RANDU ($a = 2^{16} + 3$, $m = 2^{31}$) gives $v_2^2 = 536\,805\,386$, $v_3^2 = 118$, $v_4^2 = v_5^2 = v_6^2 = 116$, indicating poor distributivity in dimensions higher than 2. This and other examples are given in Table 1 of Knuth (1981).

The most disturbing feature about the good generators shown in Table 2.1 is that it becomes very difficult to obtain good distributivity in many dimensions, with a small word length machine. For a b-bit machine, the number of k-tuples in a type A generator is 2^b. Thus k-distributivity is possible to t-bit accuracy only where $(2^t)^k \leqslant 2^b$ or $tk \leqslant b$. For a 32-bit machine, 4-distributivity is achievable (providing a good multiplier is found) to an accuracy of 8 bits. But for a 16-bit personal computer this reduces to 4 bits or an accuracy of only 1 part in 16. Herein lies the major disadvantage of

Table 2.1 l_k and the ratio l_k/l_1 for some good multipliers (from Ripley, 1983a, reproduced with permission)

Example[‡]	m	a	c	l_2 (r_2) $[m^{-1/2}]$	l_3 (r_3) $[m^{-1/3}]$	l_4 (r_4) $[m^{-1/4}]$
1	2^{59}	13^{13}	0	2.49×10^{-9} (1.23) $[2.6 \times 10^{-9}]$	2.39×10^{-6} (1.57) $[1.9 \times 10^{-6}]$	7.29×10^{-5} (1.93) $[5.1 \times 10^{-5}]$
2	2^{32}	69 069	odd	1.61×10^{-5} (1.06) $[1.5 \times 10^{-5}]$	7.25×10^{-4} (1.29) $[6.1 \times 10^{-4}]$	4.65×10^{-3} (1.30) $[3.9 \times 10^{-3}]$
3	$2^{31} - 1$	630 360 016	0	2.45×10^{-5} (1.29) $[2.2 \times 10^{-5}]$	1.61×10^{-3} (2.92) $[7.8 \times 10^{-4}]$	5.82×10^{-3} (1.64) $[4.6 \times 10^{-3}]$
4	2^{16}	293	odd	4.47×10^{-3} (1.20) $[3.9 \times 10^{-3}]$	2.64×10^{-2} (1.07) $[2.4 \times 10^{-2}]$	7.34×10^{-2} (1.45) $[6.3 \times 10^{-2}]$

[‡]Example 1 is the NAG G05CAF generator; example 2 is the result of a search for good multipliers (Marsaglia, 1972); example 3 is used in the Simscript II language; example 4 is the result of a search over multipliers 1–2000 performed by Ripley (Toothill, 1982).

the linear congruential generators. Despite all the efforts made to search for good multipliers, high-order distributivity is difficult to achieve on medium or small word length machines. Combining the output from several linear congruential generators provides some hope and this is discussed in Section 2.3.3.

2.3   Other types of generator

We observed that the linear congruential generator has a limited period, which leads to poor distributivity in higher dimensions. To increase the period, it is natural to consider *general* linear congruential generators of the type

$$X_i = (a_1 X_{i-1} + \ldots + a_p X_{i-p}) \bmod m \qquad (2.12)$$

where $p > 1$, $a_p \neq 0$. The period is the smallest positive integer λ for which

$$(X_0, \ldots, X_{p-1}) = (X_\lambda, \ldots, X_{\lambda+p-1}).$$

Since there are m^p possible p-tuples, of which one is the null vector (which leads to degeneracy of the sequence), the maximum achievable period is $m^p - 1$. Thus for $p > 1$, there is the potential for sequences of much longer periods than in the linear congruential method. In this category the simplest structure generator is the *Fibonacci* type.

2.3.1   Fibonacci generators

The use of $p = 2$, $a_1 = a_2 = 1$ leads to the Fibonacci generator

$$X_i = (X_{i-1} + X_{i-2}) \bmod m. \qquad (2.13)$$

Since no multiplications are involved this has the advantage of being fast. Wall (1960) established a number of useful results concerning its period. Maximal period is attained when $X_0 = 0$, $X_1 = 1$. Denoting $k(p)$ as the maximal period for a prime modulus p, $k(2) = 3$, $k(5) = 20$, $k(p)$ divides $p - 1$ ($p = 10x \pm 1$), $k(p)$ divides $2p + 2$ ($p = 10x \pm 3$). In the majority of cases $k(p)$ attains the values $p - 1$, $2p + 2$. For $p < 2000$, Wall gives $k(p)$ for those instances (99 of them) where it does not. If $k(p^2) \neq k(p)$ then $k(p^e) = p^{e-1} k(p)$. A computer search showed that for all $p < 10\,000$, $k(p^2) \neq k(p)$, enabling the period length $k(p^e)$ to be determined in these cases. It is not known whether the result is true for all p. A further result is that if the modulus m has the prime factorization $m = \prod p_i^{e_i}$ then $k(m)$ is the LCM of $\{k(p_i^{e_i})\}$. Also of interest is the observation that the period is often *independent* of the starting values X_0, X_1, providing X_0, X_1, and m are relatively prime. If $m = p^e$ where $p = 2$, $10x \pm 3$, then maximal period $k(p^e)$ is always obtained. If $m = 5^e$ the period is $k(5^e)/5$ or $k(5^e)$ according to whether $X_1^2 - X_0 X_1 - X_0^2 = 0 \pmod 5$

or not. If $m = p^e$ where $p \neq 5$ and $k(p) = 0 \pmod 4$ then maximal period $k(p^e)$ is always obtained.

These results are sufficient to show that the period is usually disappointingly low. For example, if m is the Mersenne prime $2^{31} - 1$, then $k(p)$ must divide $2p + 2 = 2^{32}$. So even if $k(p)$ attains the value $2p + 2$, the period is not much greater than for the equivalent linear congruential generator. Another example is $m = 2^{35} - 1 = 31 \times 71 \times 127 \times 122\,921$, for which $k(m) = \text{LCM}$ $\{k(31), k(71), k(127), k(122\,921)\}$. The first three are not included in the list of 99 exceptions, so $k(31) = 30$, $k(71) = 70$, $k(127) = 256$, $k(122\,921)$ must divide $122\,920$. Thus $k(m)$ cannot exceed the LCM of $2 \times 3 \times 5$, $2 \times 5 \times 7$, 2^8, $2^3 \times 5 \times 7 \times 439$, which is $2^8 \times 3 \times 5 \times 7 \times 439$ or $11\,800\,320$. This is much lower than for the equivalent linear congruential generator. Given that generators with a two-term recurrence relationship have the potential of achieving periods of length $m^2 - 1$, these results are disappointing. Further, Knuth (1981, p. 26) has indicated that tests show that the numbers produced are not sufficiently random. This is hardly surprising since eqn (2.13) implies that all 3-tuples (X_{i-2}, X_{i-1}, X_i) will lie on one of two planes $X_i = X_{i-1} + X_{i-2}$ or $X_i = X_{i-1} + X_{i-2} - m$. It would seem that the only way to salvage some form of randomness from such generators (which have the considerable advantage of being very fast) is to break up the undesirable lattice structure. In this connection Gebhardt (1967) has reported satisfactory results when the output from a Fibonacci generator is *shuffled*. Methods of shuffling are described in Section 2.3.3.

2.3.2   Tauseworthe generators

Returning to the general linear congruential generator (2.12), consider the case m prime. The generator has maximal period $m^p - 1$ when the constants $\{a_i\}$ are chosen so that the polynomial

$$x^p - a_1 x^{p-1} \dots - a_p$$

is a *primitive polynomial modulo* m, (Knuth, 1981, p. 28). The case which has attracted most interest is $m = 2$, leading to a sequence of bits. The modulo 2 process allows us to restrict $\{a_i\}$ to 0 or 1. The only primitive binomial modulo 2 is $1 + x$, for which $p = 1$, leading to a maximal period of 1. This is of no practical use, so *trinomials* are usually used. These are of the form

$$x^p + x^q + 1 \tag{2.14}$$

where $p > q$. Zierler and Brillhart (1968) give all primitive trinomials modulo 2, up to degree $p = 1000$. Using the trinomial (2.14) the non-zero multipliers in the right-hand side of eqn (2.12) are a_{p-q} and a_p, so the recurrence becomes

$$X_i = X_{i-p} \oplus X_{i-(p-q)} \tag{2.15}$$

where $\oplus$ is addition modulo 2. The operator $\oplus$ is the 'exclusive-or' operator defined by $0 \oplus 0 = 1 \oplus 1 = 0, 0 \oplus 1 = 1 \oplus 0 = 1$.

Tauseworthe or feedback shift register generators are named after Tauseworthe (1965) who took the binary bit generator (eqn 2.15) and assembled groups of L adjacent bits ($L \leqslant p$) to obtain numbers or words to L-bit accuracy. The leading bits in successive words are spaced Q bits apart where $Q \geqslant L$. If $Q = L$, this is known as an L-wise decimation of $\{X_i\}$. A decimation is *proper* if Q is relatively prime to $2^p - 1$. In this case the period of the L-bit words is $2^p - 1$. Tauseworthe showed that for proper decimations, the mean, variance, and correlation up to lags $(2^p - 1 - L)/Q$ are in accord with what might be expected of a random sequence. Further, the words are k-distributed for $k \leqslant \langle p/Q \rangle$. The latter is an attractive feature when compared with linear congruential generators, since p is not limited by the word length of the machine.

For example, a primitive trinomial of degree 127 is $x^{127} + x^{30} + 1$. If we choose $Q = L = 30$ (which is relatively prime to $2^{127} - 1$) then a maximal period of $2^{127} - 1$ 30-bit words are obtained, which are k-distributed for $k \leqslant \langle 127/30 \rangle = 4$. Alternatively, choosing $Q = L = 15$, 15-bit words of period $2^{127} - 1$ are obtained, which are k-distributed for $k \leqslant \langle 127/15 \rangle = 8$. The higher-dimensional distributivity of these generators arises because of the *repeatibility* of L-bit words within the cycle. Unlike the linear congruential method, the generator is hardly limited by the word length of the computer. All that is required is that L not exceed this word length. Thus high-degree primitive trinomials can give very high-dimensional distributivity, and if carefully designed they are eminently suitable for small word length machines, which are now common in the form of personal/microcomputers.

Toothill *et al.* (1971) investigated an additional property of the sequence: the pattern of runs up and down (see Section 2.4). They found that if q is close to 0, p, or $p/2$, the generator can fail this test. Toothill *et al.* (1973) introduced the concept of *asymptotic randomness*. Suppose numbers to t-bit accuracy ($t \leqslant L$) are required. For k-distributivity, the period length must be at least as large as the number of t-bit k-tuples $(2^t)^k$. Thus $2^p - 1 \geqslant 2^{kt}$. Considering the very large size of the period, -1 may be neglected and so k-distributivity is potentially achievable for $k \leqslant \langle p/t \rangle$. Toothill *et al.* argue that even when this maximal distributivity is attained, some generators will be better than others. To find the good ones, they further insist that the generator be (h, k)-distributed (see Section 2.1) for all h and all $k \leqslant \langle p/t \rangle$. Such a sequence is said to be asymptotically random. As noted previously, (h, k)-distributivity is a much stronger condition, which includes the requirement that non-overlapping t-bit k-tuples are k-distributed. It transpires that a necessary condition for (h, k)-distributivity is that the period $2^p - 1$ be prime. Interest focuses, therefore, on primitive trinomials whose degree is a Mersenne exponent. Trinomials for the first 23 such exponents have been listed by Zierler (1969). An example of a generator giving asymptotic randomness is $p = 607$, $q = 273$ or 334, $Q = 512$, $L = 23$. If 23-bit accuracy is required the sequence is (h, k)-distributed for $k \leqslant \langle 607/23 \rangle = 26$. A further example is $p = 607$, $q = 502$, $Q = L = 15$. For 15-bit accuracy this is (h, k)-distributed for $k \leqslant \langle 607/15 \rangle = 40$. Note that these generators give an astronomic period of $2^{607} - 1 = 5.31 \times 10^{182}$.

Lewis and Payne (1973) form L-bit words not by grouping adjacent bits, but by introducing a *delay* between the bits. Under specified conditions, maximal period length of $2^p - 1$ can be achieved for these words. They show that the mean and variance are theoretically correct, while the serial correlations up to lags 50 are acceptably low when a constant delay between bits of order $100p$ is used. The sequence has potential k-distributivity for $k \leqslant \langle p/L \rangle$ but little is said about what is actually achieved in this respect.

In conclusion, Tauseworthe generators offer the experimenter the potential for considerable advantages over linear congruential methods (mainly higher-dimensional distributivity on small word length machines). However, until knowledge on the theoretical properties of such generators reaches the level that it has reached for linear congruential ones, there is the danger that Tauseworthe generators with particular parameter values may yield highly non-random results. At present, perhaps the safest strategy is to use the particular examples suggested by Toothill *et al.*

2.3.3   Shuffled and portable generators

Given the inevitable dependencies that will exist in a 'random' sequence, it seems natural that one should try to *shuffle* the output of a sequence, in the hope that this will make it more random. Examples of such approaches are Maclaren and Marsaglia (1965) and Bays and Durham (1976), and methods based on these are given in Algorithms 2.2 and 2.3, respectively.

Algorithm 2.2 uses two sequences, R_1 representing the next random number in the first sequence, R_2 the next in the second sequence. Initially the first k numbers of the first sequence are put into a table $T_0, \ldots, T_{k-1}$. In step

Algorithm 2.2

Input $[R_1, R_2]$ Saved $[T_0, \ldots, T_{k-1}]$
Output $[R]$ Parameter $[k]$.

1   Data $T_0, \ldots, T_{k-1}$.
2   $n := \langle R_2 k \rangle$
3   $R := T_n$.
4   $T_n := R_1$.
5   Exit.

Algorithm 2.3

Input $[R_1]$ Saved $[T_0, \ldots, T_{k-1}, Z]$
Output $[R]$ Parameter $[k]$

1   Data $T_0, \ldots, T_{k-1}, Z$.
2   $n := \langle Zk \rangle$
3   $Z := T_n, T_n := R_1$.
4   $R := Z$.
5   Exit.

2, R_2 is used to determine a position, n, within the table, and T_n is then delivered as the output random number. The nth position within the table is replenished with R_1. The period of the output sequence $\{R\}$ is never greater than the LCM of the periods of $\{R_1\}$ and $\{R_2\}$ (see Knuth, 1981, p. 35), so the period is not improved for maximal period linear congruential generators, modulus 2^b. However, the main benefit is that the procedure goes some way to break up the regular lattice structure for linear congruential generators.

Algorithm 2.3 uses only one sequence, R_1 representing the next number within that sequence. Initially the first k numbers in the sequence are put into a table $T_0, \ldots, T_{k-1}$. The $(k+1)$th number is assigned to a variable Z. Z is used to determine a position, n, within the table. T_n is then assigned to Z, which is delivered. Meanwhile R_1 is used to replenish T_n. Since only one stream is required, the method is faster than Maclaren and Marsaglia's, and Bays and Durham indicate that the period may also be superior. Other methods of modifying output for generators are suggested by Greenwood (1976), Salfi (1974), Westlake (1967), and Kral (1972).

One of the reasons for using pseudo, rather than true random numbers is the capability to reproduce sequences. This is particularly important when verifying results produced by other experimenters. Most generators are coded in a low-level language and the implementation often depends upon the hardware, although we note that the latter is far less crucial for Tauseworthe generators. Portable generators are usually written in a general-purpose language such as FORTRAN and can be implemented independently of the machine hardware. They are therefore readily implemented and give consistent results between machines. This is obtained at the expense of a loss in speed. Examples of such generators are given by Schrage (1979) and Wichmann and Hill (1982, 1984). The latter method is a composite generator summing the output from three prime modulus linear congruential generators, and then taking the fractional part. (The fractional part of the sum of several $U(0, 1)$ random variables is also $U(0, 1)$.) The period of the generator is 6.95×10^{12}, which is the LCM of the periods of the three generators, namely 30268, 30306, and 30322. This gives the *raison d'être* for this and certain shuffled generators. High-order distributivity is possible using several linear congruential generators, even on small word length machines.

2.4   Empirical tests

Empirical tests take a sample of output from the generator and attempt to identify deviations from randomness. An infinite number of such deviations could be tested for, but we confine ourselves to detecting non-uniformity and specific forms of dependency between the numbers in the stream. At the time of writing, one of the advantages of empirical tests over theoretical tests is that few results concerning the latter are applicable to *local* segments of the stream, whereas any segment may be chosen for empirical testing. Thus, while theoretical testing is necessary and useful for detecting *global* non-randomness, empirical testing is often better for establishing local randomness. Note that empirical testing of the whole cycle of many good generators is not a practicable possibility. The length of the cycle would ensure that the computer was busy generating random numbers and doing nothing else for its entire lifetime. We now consider a number of empirical tests.

2.4.1   Frequency test

This is designed to check the uniformity of numbers within $[0, 1)$. Consider a sample of n identically and independently distributed random variables $R_1, \ldots, R_n$. We wish to establish whether the parent population is $U[0, 1)$, which is the null hypothesis. Two possible approaches use the chi-squared and Kolmogorov–Smirnov test statistics, respectively.

Under the former, the domain $[0, 1)$ is divided into k equal parts. Let f_i denote the number of observations falling into the ith such part. Under the null hypothesis the expected number falling into the ith part is n/k, and the test statistic

$$X^2 = \sum_{i=1}^{k} \frac{(f_i - (n/k))^2}{n/k}$$

is asymptotically distributed as chi-squared with $k-1$ degrees of freedom. High values of X^2 support rejection of the null hypothesis. Morgan (1984, p. 141) states that chi-squared tests of randomness are often two-tailed, low values of X^2 indicating too good a fit. Care needs to be taken in this respect. If the entire sequence of a full period type A linear congruential generator is taken, then X^2 is zero! The assumption of independence is important, which means that before implementing the test, the presence of obvious forms of dependency should be ruled out. The test is only exact asymptotically. Further, the division of the domain into k intervals loses information, with a consequent reduction in the power to detect alternative hypotheses.

The Kolmogorov–Smirnov test is an exact one for any sample size n, and compares the empirical cumulative distribution function (c.d.f.) $\hat{F}(\cdot)$ with a theoretical c.d.f. $F(\cdot)$. Under the null hypothesis $F(x) = x$, $0 \leqslant x < 1$. To obtain the empirical c.d.f. rank the observations so that $R^{(1)} \leqslant R^{(2)} \ldots \leqslant R^{(n)}$. Then

$$\hat{F}(x) = \frac{i}{n} \qquad R^{(i)} \leqslant x < R^{(i+1)},$$

where $R^{(0)} \equiv 0$ and $R^{(n+1)} \equiv 1$. The test statistic measures the size of the largest difference between these two, that is

$$D_n = \max_{0 \leqslant x < 1} |\hat{F}(x) - F(x)|.$$

D_n follows the one-sample Kolmogorov–Smirnov distribution, which has been tabulated for various n (see, for example, Lindley and Scott, 1984). Under the null hypothesis, large values of

$$D_n = \max_{0 \leqslant x < 1} |\hat{F}(x) - x|$$

indicate a poor fit and hence support rejection of the null hypothesis. For significance at $100\alpha\%$, the test statistic value D_n is compared with the $100(1 - \alpha)\%$ point of the distribution.

Rather than take just one sample of size nk, say, we could consider taking k samples of size n. A test based on the former will tend to give a more global measure of randomness, whereas the latter may be expected to detect local non-randomness, which may go undetected in the former case (see, for example, Knuth, 1981, p. 49). For k samples of size n, we obtain k test statistic values $D_{n,1}, \ldots, D_{n,k}$. Rank these so that $D_n^{(1)} \leqslant \ldots \leqslant D_n^{(k)}$, and construct an empirical c.d.f.

$$\hat{G}_n(z) = \frac{i}{k} \qquad D_n^{(i)} \leqslant z < D_n^{(i+1)},$$

where $D^{(0)} \equiv 0$ and $D^{(k+1)} \equiv 1$. The fit between $\hat{G}_n(z)$ and the theoretical c.d.f. $G_n(z)$ of $\{D_n\}$ may be examined through a further Kolmogorov–Smirnov test statistic

$$\Delta_{n,k} = \max_{0 \leqslant z \leqslant 1} |\hat{G}_n(z) - G_n(z)|.$$

In practice the sample size n will be chosen to be large ($\gtrsim 500$) and so the known asymptotic distribution of $n^{\frac{1}{2}} D_n$ may be used with great accuracy. This gives

$$G_n(z) = 1 - 2 \sum_{r=1}^{\infty} (-1)^{r-1} e^{-2nr^2 z^2}. \qquad (2.16)$$

Using this, $\Delta_{n,k}$ can easily be evaluated through a computer program. Only

Table 2.2 Results of tests on uniformity using Kolmogorov–Smirnov statistics

n	k	$\Delta_{n,k}$	Critical value at 5%
20	500	0.182	0.294
10	1000	0.409	0.409
5	2000	0.453	0.563
2	5000	0.741	0.842
1	10000	0.873	0.975
25	1000	0.169	0.264
12	2000	0.219	0.375
5	5000	0.288	0.563
3	10000	0.412	0.708

the first few terms in (2.16) need be taken, as the series converges rapidly. Table 2.2 shows values of $\Delta_{n,k}$ for two segments of output produced by the FORTRAN random number function RAN on the DEC-20 computer. This function is based upon a prime modulus generator with $m = 2^{31} - 1$ and $a = 630\,360\,016$. The results above the dashed line relate to a segment of length 10 000, where the aim is to detect local non-uniformity over parts having lengths ranging from 500 to 10 000. A longer segment of length 30 000 is considered below the dashed line. In neither case is there significant evidence of local or global deviations from uniformity.

The major advantages of the Kolmogorov–Smirnov test over the chi-squared test are that it is exact for all n, and that information is not lost by the grouping of observations. Massey (1951) has shown that for specific alternative hypotheses the power of the Kolmogorov–Smirnov test is substantially higher than that for the corresponding chi-squared test. One difficulty in implementing the Kolmogorov–Smirnov test is the requirement to sort n data items. The associated computation time is $O(n \log n)$. Fortunately Gonzalez *et al.* (1977) have devised a method which eliminates the sort and has computation time $O(n)$.

2.4.2   Serial test

A necessary condition for a stream to be random is that the *non-overlapping* 2-tuples $\{(R_{2j}, R_{2j+1})\}$ are uniformly distributed over $[0, 1)^2$. If we take a stream of length $2n$ giving n such 2-tuples, and divide the unit square into k^2 sub-squares, each of area $1/k^2$, then the expected frequency of 2-tuples falling into the ith sub-square is n/k^2. Denoting the observed frequency by f_i, the test statistic

$$X^2 = \sum_{i=1}^{k^2} \frac{(f_i - n/k^2)^2}{n/k^2}$$

is asymptotically chi-squared with $k^2 - 1$ degrees of freedom. In the case of p-tuples the statistic is chi-squared with $k^p - 1$ degrees of freedom. The p-tuples should be non-overlapping to preserve the independence required for the usual form of the chi-squared test. For k prime, Good (1953) has devised a form of the test for *overlapping* p-tuples. The serial test is an empirical analogue of the spectral and lattice tests, but note that the latter are usually implemented with overlapping p-tuples. Large values of X^2 are supportive of a non-uniformity of p-tuples through the p-dimensional hypercube. In practice the test is difficult to implement for p much larger than 2. For example, if $p = 4$ and $k = 100$, then 10^8 sub-cubes are required, causing a storage problem. Further, the chi-squared test, being asymptotic, would require at least 5×10^8 pairs, which is approximately 20 per cent of the cycle length of a 32-bit full period linear congruential generator. Such a test would give little information on local non-randomness. Although some global evidence would be forthcoming, the spectral and lattice tests would provide this far more easily for type A, B, and C linear congruential generators. In fact Knuth (1981, pp. 105–9) indicates that the serial test applied to a sufficiently long segment will always be passed by any generator passing the spectral test. The spectral test is stronger because it is rotationally invariant, unlike the serial test.

2.4.3 Gap test

In this test we observe the 'gap' between successive occurrences of random numbers lying in a specified range (R_L, R_U). Under the assumption of randomness, the gap length is geometrically distributed with probability mass function (p.m.f.)

$$p_i = \rho^i(1 - \rho)$$

where $\rho = R_U - R_L$. A predetermined number n of such gaps are generated. Let f_i $(i = 0, \ldots, k-2)$ denote the frequency of occurrence of gaps of length i, and f_{k-1} that of gap lengths greater than $k-2$. Redefining p_{k-1} as the probability of a gap length $k-1$ or larger, the test statistic

$$X^2 = \sum_{i=0}^{k-1} \frac{(f_i - np_i)^2}{np_i}$$

where $p_{k-1} = \rho^{k-1}$, is distributed asymptotically as a chi-squared with $k-1$ degrees of freedom. In implementing this test it is important to specify n in advance, rather than the length of the stream to be generated.

2.4.4 Poker test

This is a much-quoted test for sequences of digits. n groups of five digits are considered, and each group placed into one of seven categories defined below:

> Digits   are all different,
> have one pair,
> have two pairs,
> have one triple and two distinct,
> have one triple and one pair,
> have four identical,
> are all identical.

Knuth (1981, p. 62) suggests a version of the test which is simpler to implement and is almost as good. In this the number of distinct digits in a group is observed, giving five categories.

To determine p_r, the probability (assuming randomness) that a group contains exactly r distinct digits, note that each 5-tuple can take one of d^5 equally likely values, where each digit $\in[0, d-1]$. The number of ways in which a 5-tuple can have exactly r distinct digits is the number of ways of partitioning five objects into r parts multiplied by the number of orderings of r objects from d. The latter is $d(d-1)\ldots(d-r+1)$. The former is $S(5, r)$ where $S(k, r)$ denotes the number of ways of partitioning k objects into r parts. $\{S(k, r)\}$ are the Stirling numbers of the second kind (see, for example, Riordan, 1958, p. 33). Clearly $S(k, 1) = S(k, k) = 1$. Other values are determined through the recurrence

$$S(k, r) = rS(k-1, r) + S(k-1, r-1). \tag{2.17}$$

Some low-order values are shown in Table 2.3. Thus we have

$$p_r = \frac{S(5, r)d(d-1)\ldots(d-r+1)}{d^5}.$$

Considering decimal digits 0–9, $\{10^4 p_r\}$ are readily computed as 1, 135, 1800, 5040, and 3024, respectively.

Table 2.3 Low-order Stirling numbers of the second kind, $S(k, r)$

k/r	1	2	3	4	5	6	7	8	9	10
1	1									
2	1	1								
3	1	3	1							
4	1	7	6	1						
5	1	15	25	10	1					
6	1	31	90	65	15	1				
7	1	63	301	350	140	21	1			
8	1	127	966	1701	1050	266	28	1		
9	1	255	3025	7770	6951	2646	462	36	1	
10	1	511	9330	34105	42525	22827	5880	750	45	1

The test can be extended to deal with groups of k digits at a time. As in previous tests the number of groups falling into each category is observed and compared with the critical values from a chi-squared distribution with $k-1$ degrees of freedom. The test may also be used on the random numbers themselves, rather than the digits, by dividing $[0, 1)$ into k disjoint sub-intervals.

2.4.5   Coupon collector's test

Consider again a sequence of random digits uniformly distributed in $[0, d-1]$. Let C_d be the number of digits required before the 'full set' of d digits is first obtained. Under randomness,

$$\text{prob}(C_d \leqslant k) = \text{prob(all } d \text{ digits are represented in a}$$
$$\text{sequence of } k \text{ consecutive digits)}$$
$$= \frac{S(k, d)d!}{d^k} \tag{2.18}$$

for $k \geqslant d$, using the arguments developed for the poker test. The probability p_k that the full set is obtained after exactly k digits is

$$p_k = \frac{S(k, d)d!}{d^k} - \frac{S(k-1, d)d!}{d^{k-1}}$$

which gives

$$p_k = \frac{d!\, S(k-1, d-1)}{d^k} \qquad (k \geqslant d), \tag{2.19}$$

on using eqn (2.17). Note that in the case of binary digits $d = 2$ and $p_k = 2^{1-k}$.

To implement the test, n consecutive 'full sets' are generated. Let f_k denote the frequency of occurrence of $C_d = k$ for $k = d, \ldots, d+m-2$, respectively, and f_{d+m-1} the frequency of occurrence of $C_d \geqslant d+m-1$. The observed frequencies $\{f_k\}$ are compared with the expected frequencies (computed from eqns (2.18) and (2.19)) using a chi-squared test with $m-1$ degrees of freedom).

2.4.6   The collision test

We noted previously that the serial test is difficult to apply in many dimensions. For example the distribution of 4-tuples through 10^8 sub-cubes or 'cells' requires at least 5×10^8 observations. The collision test explores higher-dimensional randomness using far fewer observations.

Suppose we wish to explore the distribution of n observations through m cells where $m \gg n$. Most cells will be empty, and this is why the usual chi-squared serial test will not work. Instead of recording the frequency of observations in each cell (which will be low) record the total number of 'collisions' occurring. A collision occurs if an observation falls into a cell which is already occupied. Thus the number of collisions C is determined by the number of cells containing at least two observations or

$$C = n - \text{number of occupied cells.}$$

To determine the p.m.f. of C, we observe that the total number of ways in which n observations may be distributed through m cells is m^n. The number of these corresponding to exactly c collisions is the number of ways of partitioning n observations into $n - c$ parts, multiplied by the number of orderings of $n - c$ objects from m objects. Thus the probability of exactly c collisions is

$$\text{prob}(C = c) = \frac{S(n, \, n - c)m \ldots (m - n + c + 1)}{m^n}.$$

Suppose we wish to implement the test for the 4-tuples discussed previously. Although 10^8 storage locations are required to determine whether cells are occupied or not, these locations need only record bits (0 for unoccupied, 1 for occupied), and so the memory requirements are just manageable on most machines. The number of collisions recorded is compared with the percentage points of the distribution. Although m and n are large the expected number of collisions is small ($\sim n^2/2m = 50$). Knuth (1981, pp. 69–70) gives an algorithm for computing the percentage points.

2.4.7   Runs test

Consider the following number sequence:

0.415, 0.919, | 0.800, | 0.677, | 0.594, | 0.059, 0.444, 0.618, | 0.429, 0.779, 0.788, 0.878, 0.914, 0.952, ...

The demarcation lines show the end of an ascending sequence of numbers. Thus the length of ascending runs is 2, 1, 1, 1, 3, ... The runs test compares the observed frequency of different run lengths with the expected frequency. Levene and Wolfowitz (1944) showed that the expected frequency of ascending runs of length k is

$$e_k = \frac{(k^2 + k - 1)(n - k - 1)}{(k + 2)!},$$

where n is the length of the sequence.

Unfortunately the usual chi-squared test statistic cannot be used, due to the dependencies between successive run lengths. A rather complicated test statistic utilizing the approximation

$$e_k \simeq \frac{(k^2 + k - 1)n}{(k + 2)!}$$

is given in Knuth (1981, p. 65).

A far more practicable approach is recommended by Knuth. The number immediately following the end of a run is deleted. Thus the sequence above becomes

0.415, 0.919, | 0.677, | 0.059, 0.444, 0.618, | 0.779, 0.788, 0.878, 0.914, 0.952, ...

Successive run lengths are now independent, and this test reduces to the usual chi-squared form. Specifically the probability of obtaining a run of length k is

$$\text{prob}(R_1 < \ldots < R_k \geqslant R_{k+1}) = \int_0^1 dr_{k+1} \int_{r_{k+1}}^1 dr_k$$

$$\int_0^{r_k} dr_{k-1} \ldots \int_0^{r_2} dr_1 = \frac{k}{(k+1)!}$$

and the probability that a run of length at least m is obtained is $1/m!$. From a predetermined number v of ascending runs the observed frequencies $\{f_i\}$, $i = 1, \ldots, m-1$ of run length i and f_m of run lengths of at least m are obtained. The test statistic is

$$X^2 = \sum_{k=1}^{m-1} \frac{\{f_i - (vk/(k+1)!)\}^2}{vk/(k+1)!} + \frac{\{f_m - (v/m!)\}^2}{v/m!}$$

which, assuming randomness, is distributed as chi-squared with $m-1$ degrees of freedom.

2.4.8   Linear dependence

One way to test for dependencies between numbers in a sequence is to restrict such examination to linear dependence between observations which are separated by k numbers. Given a realization of n random numbers $R_1, \ldots, R_n$, the sample covariance of lag k is

$$C_k = (n-k)^{-1} \sum_{i=1}^{n-k} (R_i - \tfrac{1}{2})(R_{i+k} - \tfrac{1}{2}).$$

Under randomness $E(C_k) = 0$ and $V(C_k) = \{144(n-k)\}^{-1}$. For $n \gg k$, the distribution is approximately Normal, so Fishman (1978, p. 383) recommends that the statistic $Z_k = 12C_k\sqrt{(n-k)}$ be compared with critical points of the normal distribution.

To illustrate the method, two sequences of length 10 000 were generated using the DEC-20 FORTRAN random number generator described in Section 2.4.1. Table 2.4 shows the values of the associated statistics, $Z_k^{(1)}$ and $Z_k^{(2)}$, for lags ranging from 1 to 25. For these lags, there appears to be little evidence of linear dependence, apart perhaps from $k = 25$, on the second sequence, although Z_{25} is well within limits for the first sequence.

Table 2.4 $Z_k^{(1)}$ and $Z_k^{(2)}$ for lags up to 25 for two sequences of length 10 000

k	1	2	3	4	5	6	7	8	9	10
$Z_k^{(1)}$	0.4115	-1.0830	-0.0082	-0.0153	0.3110	0.7388	-1.6881	-0.3439	-0.8078	-0.5835
$Z_k^{(2)}$	1.0760	-0.4122	-0.4769	-0.0346	1.5953	0.4826	1.4586	-2.0082	0.9043	0.8764

k	11	12	13	14	15	16	17	18	19	20
$Z_k^{(1)}$	-0.0573	-0.1065	0.7687	0.3379	0.6047	-0.3965	-0.5320	1.1159	-0.7190	0.6294
$Z_k^{(2)}$	0.5963	-0.0197	-0.9510	0.0229	-0.9497	1.0508	0.2729	-1.0948	-0.3868	2.2081

k	21	22	23	24	25
$Z_k^{(1)}$	-0.0982	-1.0459	0.5069	-0.2890	0.8461
$Z_k^{(2)}$	-1.5701	-0.4373	-1.0468	0.7973	-3.0566

A related test is to examine the serial correlation, rather than the covariance (Knuth, 1981, pp. 70–1). Since the expectation of the sample serial correlation is not exactly zero, the test based on Z_k is easier to implement.

2.4.9   Concluding remarks

There is of course no limit to the number and diversity of empirical tests that can be applied. Amongst those we have not described are the d^2-test (Gruenberger and Mark, 1951), the *maximum-of-t-test* and the *permutation test* (Knuth, 1981, pp. 64, 68). Knuth cites the runs, collision, and maximum-of-t-test as often being more sensitive, in the sense that defective generators passing the other tests frequently fail one or more of these. The history of random number testing begins with Kendall and Babbington Smith (1938, 1939) who considered the frequency, serial, gap, and poker tests applied to random digits rather than random numbers in [0, 1). They applied the serial test to *overlapping* pairs, but used the incorrect chi-squared statistic, on the assumption of independence of pairs. The idea of taking several sampling statistics from different segments of a sequence was discussed under the frequency test. This practice can and should be applied to all the empirical tests. It should lead to a more sensitive procedure for detecting local non-randomness. In this connection a plot of ranked chi-squared statistic values against the expected order statistics (Wilk *et al.*, 1962) provides a rough and ready assessment when the number of statistics is reasonably large. Atkinson (1977a) suggests that the distribution of a sample of k-tuples through $[0, 1)^k$ be investigated using a technique developed by Ripley for assessing randomness in spatial point processes. On a pessimistic note, Atkinson in the same paper takes the view that 'empirical tests based on sampling are rarely powerful enough to detect any but the worst generators'. This is not too serious for linear congruential generators for which adequate theoretical tests are now available. It is serious, however, for Tauseworthe generators, which no doubt would be used more if their theoretical properties were better understood.

Finally, the reader may be rightly concerned that the sequences considered in this chapter produce numbers in the interval [0, 1). There is a chance, albeit very remote for long-period generators, of outputting a value of zero, which would result in overflow if dividing by or taking the logarithm of the random number. Such an event is impossible with a prime modulus generator, since, assuming a non-zero seed is used, zero cannot appear in the sequence. It is a possibility with a mixed linear congruential generator. For this reason, we will assume henceforth that the random number sequence used in the algorithms are either prime modulus or have had the zero removed from the sequence. Thus a random number is taken to be uniformly distributed in the interval (0, 1).

3
General methods for generating random variates from univariate distributions

---

We now turn to a discussion of general methods and principles for generating random variates from univariate distributions, both continuous and discrete. Amongst these general methods we include inversion of the distribution function, the composition method, stochastic model methods, various rejection methods, Alias rejection, and methods suitable for discrete empirical distributions. Where appropriate, theoretical measures of performance (see Section 1.5.3) will be derived. The general methods will be illustrated by reference to certain distributions, but it should be stressed that such illustrations do not necessarily indicate the best method for the distribution. Such methods will be discussed in Chapters 4 and 5.

3.1 Inversion method

Given that we wish to generate random variates with a p.d.f. $f_X(\cdot)$, and c.d.f. $F_X(\cdot)$, we can show that

$$X = F_X^{-1}(R) \tag{3.1}$$

has the required distribution, where $R \sim U(0, 1)$. For, under the stated transformation,

$$P(X \leqslant x) = P(F_X^{-1}(R) \leqslant x)$$
$$= P(R \leqslant F_X(x))$$
$$= F_X(x).$$

If X is a *discrete* random variable, defined on the ordered values $S \equiv \{x^{(1)}, x^{(2)}, \ldots\}$, then variates may be generated by finding the smallest values of $X \in S$, such that

$$F_X(X) \geqslant R. \tag{3.2}$$

To illustrate the inversion method for a continuous random variable, consider generation from the three-parameter Weibull distribution with p.d.f.,

$$f_X(x) = \begin{cases} c(x-a)^{c-1} e^{-((x-a)/b)^c} / b^c & (x \geqslant a) \\ 0 & (x < a) \end{cases}$$

where $b > 0$ and $c > 0$. The c.d.f. is

$$F_X(x) = \begin{cases} 1 - e^{-((x-a)/b)^c} & (x \geqslant a) \\ 0 & (x < a). \end{cases}$$

Using (3.1) we obtain

$$X = F_X^{-1}(R) = a + b[-\ln(1-R)]^{1/c},$$

or, since R is identically distributed to $(1-R)$,

$$X = a + b[-\ln R]^{1/c}. \tag{3.3}$$

As an example of discrete random variate generation, consider the geometric distribution with p.m.f.

$$f_X(x) = (1-p)^{x-1} p \qquad (x = 1, 2, \ldots),$$

where $0 < p < 1$. The c.d.f. is

$$F_X(x) = 1 - (1-p)^x,$$

and using eqn (3.2) we must find the smallest integer X satisfying

$$1 - (1-p)^X \geqslant R,$$

or

$$\ln(1-R) \geqslant X \ln(1-p).$$

As before $(1-R)$ may be replaced by R to give

$$X = 1 + \langle \ln R / \ln(1-p) \rangle.$$

The suitability of the inversion method depends mainly on whether the c.d.f. can be inverted analytically. If it can, it is often a good method to use, since it requires only one random number per variate generated and has low storage requirements since no look-up table for $F_X(\cdot)$ is required. The marginal generation time is unlikely to be excessive unless $F_X^{-1}(R)$ contains several time-consuming functions such as logarithmic, trigonometric or exponential evaluations. The only circumstances in which set-up time is likely to be high is where the p.d.f. is of the form $f_X(x) = Ag(x)$, where $g(x)$ is known, but A has to be computed numerically.

If the c.d.f. cannot be inverted analytically, then, for a continuous random variable, one possibility is to find an approximation to the inverse c.d.f. For example, Page (1977) uses the following approximation to the standard normal c.d.f.:

$$\Phi^{-1}(p) \simeq u - \frac{1}{3a_2 u} \qquad (p > 0.5),$$

where

$$u^3 = \{y + (y^2 + 4/[27a_2])^{\frac{1}{2}}\}/2a_2,$$
$$y = \ln[p/(1-p)]/2a_1,$$
$$a_1 = (2/\pi)^{\frac{1}{2}},$$
$$a_2 = 0.044715.$$

In the case of the normal distribution, it is perhaps difficult to imagine circumstances where such a method will be used, since fast and exact procedures are available using other methods (see Section 4.2). As Hoaglin (1976) says, 'In general we should now regard approximate algorithms as a last resort, to be turned to only when exact ones are unavailable or hopelessly inefficient.' Despite Hoaglin's comment we may identify two sets of circumstances where an approximate method might be used even though an exact method exists. Firstly, if an approximation is extremely good, the resulting error may be not much greater than those errors arising from either the finite word length of the computer or approximations used in evaluating transcendental functions which happen to appear in the exact method.

The second situation arises when we wish to use a variance reduction technique. The *primary* and *antithetic* random variables $X = F_X^{-1}(R)$ and $Y = F_X^{-1}(1 - R)$ will frequently display negative correlation (the correlation is -1 if the distribution is symmetric. This is useful from an experimental design viewpoint, since variance reduction (See Section 1.1) can be achieved by 'pooling' the two responses—the method of *antithetic variates*. Some of the other methods for generating random variates, in particular rejection methods, are unlikely to yield the same degree of negative correlation, because of the difficulty in ensuring that the same random numbers are used to generate both the primary and antithetic variates. However, this remark is of little consequence for the normal or indeed any symmetric distribution, since a perfectly negatively correlated pair (X, Y) may be obtained by setting

$$Y = \mu - (X - \mu) = 2\mu - X. \tag{3.4}$$

To utilize (3.4) it is of course essential that X can be replicated on the antithetic run. This can be achieved by reserving a random number stream for the X values, or alternatively by storing the generated X values for use in calculating Y values during the antithetic run.

The remark is of relevance to certain approximate methods available for asymmetric distributions. For example, although many exact gamma generation algorithms have been devised using rejection methods, the rejection step is likely to reduce the induced negative correlation between primary and antithetic variates. One method, based on approximating the inverse c.d.f., is due to Tadikamalla and Ramberg (1975). A gamma distribution with mean μ^* and variance σ^{*2} is approximated by a four-parameter distribution having c.d.f.

$$F_X(x) = 1 - \left\{ 1 + \left[\mu + \left(\frac{x - \mu^*}{\sigma^*} \right) \sigma \right]^c \right\}^{-k}, \tag{3.5}$$

where $x \geqslant \mu^* - \mu\sigma^*/\sigma$, and μ, σ, $c(>0)$, $k(>0)$ are parameters chosen so that the first four standardized moments about the mean are identical to the corresponding moments of the gamma distribution. Tadikamalla and Ramberg produce regression equations which facilitate the fitting process. Since (3.5) can be inverted analytically, this approximate inversion method may be useful on those occasions when the generation of high-quality antithetic variates is of more importance than the generation of exact variates.

3.2 Composition method

In certain cases of interest $f_X(\cdot)$ can be expressed as a weighted sum (or probability mixture) of other p.d.f.s. If all the weights are positive, the composition method exploits this probability mixture representation. Thus

$$f_X(x) = \sum_{j=0}^{\infty} p_j f_{X_j}(x)$$

where $\{f_{X_j}(x)\}$ are p.d.f.s, and $\{p_j\}$ are non-negative constants with $\sum_{j=0}^{\infty} p_j = 1$ is such a probability mixture representation for $f_X(x)$. The $\{p_j\}$ can be thought of as forming a p.m.f. for a discrete random variable Y, with $p_j = \text{prob}(Y=j)$. The first stage of the composition method consists of generating a typical value of Y, using the $\{p_j\}$. If the particular value is denoted by j, the second stage consists of delivering a typical value of X_j from the p.d.f. $f_{X_j}(\cdot)$. When inversion has failed on $f_X(\cdot)$, it may be possible to represent it as a probability mixture, where the task of generating Y and $\{X_j\}$ values is very much easier than that of generating X values.

To illustrate the method, we consider a remark by Parker (1972, p. 21) who points to the apparent lack of methodology for generating variates from the p.d.f.

$$f_X(x) = A \sinh(xc)^{\frac{1}{2}} e^{-bx} \qquad (x \geqslant 0),$$

by an exact method. b and c are non-negative constants, and A is a constant to ensure that $f_X(x)$ integrates to 1. The distribution arises in atomic physics, X representing the energy of an emergent neutron born at fission.

By expanding $\sinh(xc)^{\frac{1}{2}}$, we have

$$f_X(x) = A \sum_{j=0}^{\infty} \frac{(xc)^{(2j+1)/2} e^{-bx}}{(2j+1)!}$$

$$= A \sum_{j=0}^{\infty} \frac{\Gamma(j+\frac{3}{2}) c^{j+\frac{1}{2}} x^{j+\frac{1}{2}} e^{-bx}}{(2j+1)! \Gamma(j+\frac{3}{2})}. \tag{3.6}$$

If we consider a random variable X_j, having a gamma density (Section 4.3) with shape parameter $(j+\frac{3}{2})$ and scale parameter b^{-1}, then eqn (3.6) may be expressed as

$$f_X(x) = A \sum_{j=0}^{\infty} \frac{\Gamma(j+\frac{3}{2}) c^{j+\frac{1}{2}}}{(2j+1)! \, b^{j+\frac{3}{2}}} \cdot f_{X_j}(x),$$

or

$$f_X(x) = A \sum_{j=0}^{\infty} \frac{\sqrt{\pi}\, c^{j+\frac{1}{2}}}{2^{2j+1} j!\, b^{j+\frac{1}{2}}} \cdot f_{X_j}(x)$$

$$= \frac{A\sqrt{(\pi c)}\, e^{c/(4b)}}{2b^{3/2}} \sum_{j=0}^{\infty} \left(\frac{c}{4b}\right)^j \frac{e^{-c/(4b)}}{j!} \cdot f_{X_j}(x)$$

$$= \sum_{j=0}^{\infty} p_j f_{X_j}(x),$$

where p_j is the probability that a Poisson variate, mean $c/(4b)$, takes a value j. This analysis gives incidentally the value of A as $2b^{\frac{3}{2}} e^{-c/(4b)}/\sqrt{(\pi c)}$. This probability mixture representation means that an exact generation method is in fact possible by setting

$$X = \chi^2_{2Y+3}/(2b),$$

where Y is a Poisson variate mean $c/(4b)$ and χ_n^2 is a chi-squared variate with n degrees of freedom. Equivalently (see Section 4.5.1) $2bX$ is a non-central chi-squared random variable, with three degrees of freedom and non-centrality parameter $c/(2b)$.

In practice the best way of generating chi-squared variates is to use a gamma routine (see Section 4.5.1). Note that an essential requirement for such a gamma generator is that it be efficient when the shape parameter changes between successive calls (since Y will vary). In this respect Algorithm G4 of Section 4.3.4 is recommended. Exact methods of Poisson variate generation are discussed in Section 5.1. Note that unless b and c change between successive calls, all that is required is a generator where the mean of the Poisson remains fixed between calls.

3.3   Stochastic model methods

In this class of methods, a process or sampling procedure which gives rise to a statistic having the required distribution is identified. The process is then simulated to give typical values of the statistic.

For example, a Poisson variate X having p.m.f.,

$$f_X(x) = \frac{\lambda^x e^{-\lambda}}{x!} \qquad (x = 0, 1, \ldots),$$

represents the number of events per unit time in a Poisson process, rate λ. The inter-event time is negative exponential, mean λ^{-1}. Thus X is the number of complete and independent negative exponential variates (mean λ^{-1}) that can be fitted into a unit time interval. Negative exponential variates may be obtained by setting $a = 0$, $b = \lambda^{-1}$ and $c = 1$ in eqn (3.3). Consequently to generate Poisson variates, given a stream of random numbers $\{R_i\}$, we require the largest integer X satisfying

$$\sum_{i=1}^{X} -\lambda^{-1} \ln R_i \leqslant 1,$$

or equivalently

$$\prod_{i=1}^{X} R_i \geqslant e^{-\lambda}.$$

Such a method requires a mean of $\lambda + 1$ random numbers per variate and so is suitable only when λ is small. More suitable methods for this distribution are discussed in Chapter 5.

As a second discrete distribution example, consider the hypergeometric distribution with p.m.f.,

$$f_X(x) = \frac{\binom{g}{x}\binom{N-g}{n-x}}{\binom{N}{n}},$$

where X represents the number of reliable items from a random sample of $n (\leqslant N)$, drawn without replacement, from a population consisting of g reliable items and $(N-g)$ defective ones. To simulate this process, note that the probability that the first draw results in a reliable item is g/N, while the future probabilities depend upon previous outcomes. For example, the (conditional) probability of the second draw resulting in a reliable item is $(g-1)/(N-1)$ or $g/(N-1)$ according to whether the first draw was reliable or defective. Hence Algorithm 3.1 is an algorithm for generating typical values. At most n random numbers are required per variate. This upper bound

Algorithm 3.1

```
Input [n, N, g]
Output [X]
1    gg:=g, NN:=N, i:=1, X:=0.
2    While gg>0 and i≤n do
3        p:=gg/NN
4        Generate R ~ U(0, 1)
5        If R≤p then
6            X:=X+1
7            gg:=gg-1
8        End If
9        i:=i+1
10       NN:=NN-1
11   End While.
12   Exit.
```

suggests that the method is inefficient for large n. Alternatives, possibly exploiting the fit between binomial and hypergeometric distributions, would then have to be investigated.

Stochastic model methods are usually easy to implement because the statistic of interest is generated in a fairly direct fashion from the process. Set-up times are usually small, and tables of constants are not often required. However the approach frequently generates more information than is required. Thus in the previous example, the sampling method gives X values not only for a hypergeometric, sample size n, but also for associated distributions with parameter values $n-1, n-2, \ldots, 1$. Where low efficiency occurs in stochastic model methods, the underlying reason is often the provision of redundant information.

3.4   Envelope rejection

A number of rejection methods exist, but the feature common to all of them is that a prospective sample variate is subjected to a random test which results either in acceptance or rejection of the sample variate. Rejection methods are a natural choice when inversion 'fails' through inability to invert the c.d.f. analytically. We consider first envelope rejection.

Suppose it is required to generate variates having a p.d.f. $f_X(\cdot)$ and c.d.f. $F_X(\cdot)$. The rejection method samples prospective variates from a different (*target*) distribution with p.d.f. $g_Y(\cdot)$ and c.d.f. $G_Y(\cdot)$, and applies a rejection/acceptance criterion, such that accepted Y have the desired distribution. The p.d.f. $g_Y(\cdot)$ should be chosen with two factors in mind. Firstly, the generation of Y values should be fast and exact. Secondly, $g_Y(\cdot)$ should imitate $f_X(\cdot)$ as closely as possible, otherwise the proportion of rejected variates becomes unacceptably high. A measure of the dissimilarity between $f_X(\cdot)$ and $g_Y(\cdot)$ is

Algorithm 3.2

Input $[f_X(\cdot), g_Y(\cdot)]$ Saved $[M]$
Output $[X]$

S1 $M = \max_{x} [f_X(x)/g_Y(x)]$

1   Generate Y from $g_Y(\cdot)$ [first stage sampling]
    and $R \sim U(0, 1)$.

2   If $R > \dfrac{1}{M} \dfrac{f_X(Y)}{g_Y(Y)}$ goto 1.

3   $X := Y$, Exit.

provided by

$$M = \max_{x} \left\{ \frac{f_X(x)}{g_Y(x)} \right\},$$

the maximization being over all x satisfying $f_X(x) > 0$.

The general sampling procedure is given in Algorithm 3.2. The validity of the algorithm is established by showing that the c.d.f. of Y conditional upon $RM < f_X(Y)/g_Y(Y)$ is $F_X(\cdot)$. Let $h_{Y,R}(y,r) = g_Y(y)$ be the joint p.d.f. of Y and R. Then the c.d.f. of accepted variates is

$$P\left(Y \leqslant x \mid R < \frac{1}{M}\frac{f_X(Y)}{g_Y(Y)}\right) = \frac{P\left(R < \frac{1}{M}\frac{f_X(Y)}{g_Y(Y)}; Y \leqslant x\right)}{P\left(R < \frac{1}{M}\frac{f_X(Y)}{g_Y(Y)}\right)}$$

$$= \frac{\displaystyle\int_{-\infty}^{x} dy \int_{0}^{\frac{1}{M}\frac{f_X(y)}{g_Y(y)}} h_{Y,R}(y,r)\, dr}{\displaystyle\int_{-\infty}^{\infty} dy \int_{0}^{\frac{1}{M}\frac{f_X(y)}{g_Y(y)}} h_{Y,R}(y,r)\, dr}$$

$$= \frac{\displaystyle\int_{-\infty}^{x} g_Y(y)\, dy \int_{0}^{\frac{1}{M}\frac{f_X(y)}{g_Y(y)}} dr}{\displaystyle\int_{-\infty}^{\infty} g_Y(y)\, dy \int_{0}^{\frac{1}{M}\frac{f_X(y)}{g_Y(y)}} dr}$$

$$= \frac{\displaystyle\int_{-\infty}^{x} g_Y(y)\left[\frac{1}{M}\frac{f_X(y)}{g_Y(y)}\right] dy}{\displaystyle\int_{-\infty}^{\infty} g_Y(y)\left[\frac{1}{M}\frac{f_X(y)}{g_Y(y)}\right] dy} = \frac{F_X(x)/M}{F_X(\infty)/M} \qquad (3.7)$$

$$= F_X(x),$$

as required. The algorithm is also valid for discrete distributions where $f_X(\cdot)$ and $g_Y(\cdot)$ are interpreted as p.m.f.s. The denominator in (3.7) indicates that the probability of acceptance (the *sampling efficiency*) is M^{-1}. Clearly $M \geqslant 1$, but a good choice of $g_Y(\cdot)$ will result in M being close to unity.

To illustrate the method, suppose it is desired to generate standard normal variates X, using variates Y from a standard Cauchy distribution. Then

$$f_X(x) = (2\pi)^{-\frac{1}{2}} e^{-\frac{1}{2}x^2},$$

and
$$g_Y(x) = [\pi(1+x^2)]^{-1}.$$
Thus
$$M = \max_x \left\{ \left(\frac{\pi}{2}\right)^{\frac{1}{2}} (1+x^2) e^{-\frac{1}{2}x^2} \right\} = \left(\frac{2\pi}{e}\right)^{\frac{1}{2}} = 1.520,$$

and the rejection condition in step 2 of Algorithm 3.2 becomes
$$R > \tfrac{1}{2}(1 + Y^2) e^{-\frac{1}{2}(Y^2 - 1)}.$$

First-stage sampling from the Cauchy distribution is conveniently performed by noting that
$$G_Y(x) = \frac{1}{2} + \frac{1}{\pi} \arctan x.$$

Given a random number R_1, inversion leads to
$$Y = \tan[\pi(R_1 - \tfrac{1}{2})].$$

Figure 3.1 shows the functions $f_X(x)$ and $Mg_Y(x)$, indicating how the latter function dominates or envelopes the former. Algorithm 3.3 illustrates the method.

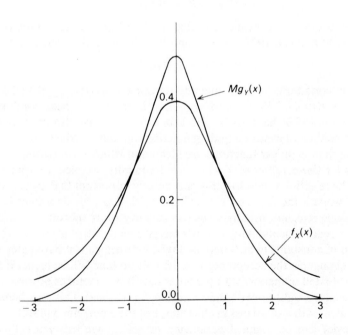

Fig. 3.1 Envelope rejection for a normal distribution with a Cauchy target distribution

Algorithm 3.3

Output $[X]$
1   Generate R, $R_1 \sim U(0, 1)$.
2   $X := \tan[\pi(R_1 - 0.5)]$.
3   If $R > 0.5 (1 + X^2) \exp[-0.5(X^2 - 1)]$ goto 1.
4   Exit.

Since $M = 1.520$, the procedure requires a mean of 3.04 random numbers per generated variate. Since the procedure also requires evaluation of exponential and tangent functions, it is unlikely to be very fast. More competitive normal procedures are discussed in Section 4.2.

Envelope rejection is a very frequently used technique. The marginal generation time depends principally upon the value of M and the complexity of the computations in the acceptance and target variate generation steps. The aim is to find a target distribution for which M is fairly close to 1, and this is often achieved through a probability mixture representation for $g_Y(x)$, with application of the composition method.

The acceptance condition may be expressed as

$$\phi(Y, R) < \psi(Y, R) \qquad (3.8)$$

where it is quicker to evaluate ψ than ϕ. The 'squeeze' technique (see, for example, Marsaglia, 1977) aims to find bounds ϕ_1 and ϕ_2 such that

$$\phi_1(Y, R) \leqslant \phi(Y, R) \leqslant \phi_2(Y, R).$$

Sufficient conditions for accepting and rejecting Y become $\phi_2(Y, R) \leqslant \psi(Y, R)$ and $\phi_1(Y, R) > \psi(Y, R)$, respectively. Only if neither of these conditions is satisfied does (3.8) have to be checked explicitly. The effectiveness of the squeeze method depends on finding functions ϕ_1 and ϕ_2 which are quicker to evaluate than ϕ (linear functions are ideal) and which form reasonably tight bounds for those values of Y which are frequently sampled. In applications where the rejection rate is low it is particularly important that ϕ_2 give a tight bound, while if the rejection rate is high ϕ_1 should provide a tight bound.

Envelope rejection usually gives rise to fairly short subroutines, with low memory requirements. Any set-up time is usually attributable to the calculation of constants (which may be saved between calls if parameter values do not change) in the acceptance test. The set-up time will be higher if M has to be evaluated numerically. In the latter case it is sometimes possible to find (analytically) a tight upper bound to $\max[f_X(x)/g_Y(x)]$. This may be used as an M value, with a small loss in efficiency, but the algorithm will still be exact. This device can be useful if parameter values change between calls of the generator, in which case each generated variate incurs a set-up time. An advantage of envelope rejection over inversion is that the p.d.f. is required to be known only up to a multiplicative constant. Consider, for example, the generation of variates from the p.d.f.

$$f_X(\theta) = Ae^{k\cos\theta} \qquad |\theta| \leqslant \pi,$$

where k is a positive constant, and A ensures that the density integrates to 1. In fact A is related to $I_0(k)$, a modified Bessel function of the first kind and order zero. Numerical inversion of the c.d.f. would require A to be evaluated numerically. In contrast A 'cancels out' in the acceptance condition of a rejection implementation. Section (4.6.3) gives a rejection method for this von Mises distribution.

3.5   Band rejection

Payne (1977) refers to this little-known method, which is a refinement of the envelope rejection method, applicable only to p.d.f.'s having a finite domain. His brief description of the method implies Algorithm 3.4 for sampling from a p.d.f. $f_X(x)$ with $0 \leqslant x \leqslant a$. Payne's original method appears to be based on a uniform target distribution. Algorithm 3.4 can also be used when $g_Y(\cdot)$ is non-uniform. In Figure 3.2, if a point (Y, R) falls beneath the lower curve, Y is delivered (primary acceptance). If this fails, the point may lie above the upper curve, in which case $a - Y$ is delivered (secondary acceptance). Only when a

Algorithm 3.4

Input $[f_X(\cdot), g_Y(\cdot)]$ Saved $[M]$
Output $[X]$

S1 $M := \max_{0 \leqslant x \leqslant a} \ [\{f_X(x) + f_X(a-x)\}/g_Y(x)]$.

1   Generate Y from $g_Y(\cdot)$
     and $R \sim U(0, 1)$.   (first-stage sampling)

2   If $R < \dfrac{1}{M} \dfrac{f_X(Y)}{g_Y(Y)}$ then

3       $X := Y$
4   Exit   (primary acceptance)

5   Else If $1 - R < \dfrac{1}{M} \dfrac{f_X(a-Y)}{g_Y(Y)}$ then

6       $X := a - Y$
7   Exit   (secondary acceptance)
8   Else
9   Goto 1   (rejection)
10 End If.

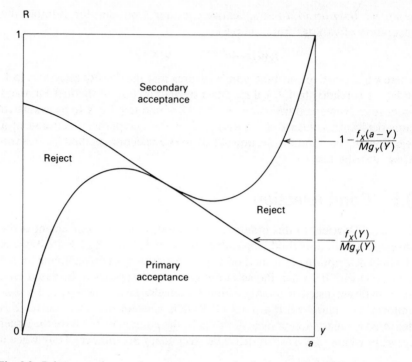

Fig. 3.2. Primary and secondary acceptance and rejection regions in band rejection method

point lies in the central region does rejection occur. The plots shown in Figure 3.2 are for functions

$$f_X(x)=\begin{cases} 1.24867xe^{-x} & (0\leqslant x\leqslant 3) \\ 0 & \text{(elsewhere)} \end{cases}$$

and

$$g_Y(x)=\begin{cases} \frac{1}{3} & (0\leqslant x\leqslant 3) \\ 0 & \text{(elsewhere).} \end{cases}$$

Note that the diagram remains unchanged if it is rotated through 180°. However, this property is lost if $g_Y(\cdot)$ is chosen to be non-uniform.

To demonstrate the validity of the algorithm we will show that the c.d.f. of X constructed in this way is $F_X(\cdot)$. Let $f_{Y,R}(y,r)=g_Y(y)$ be the joint p.d.f. of Y and R. Then,

$$P(X \leqslant x) = P\left[\left\{Y \leqslant x; R < \frac{f_X(Y)}{Mg_Y(Y)}\right\} \cup \left\{a - Y \leqslant x; 1 - R < \frac{f_X(a-Y)}{Mg_Y(Y)}\right\}\right|$$
$$R < \frac{f_X(Y)}{Mg_Y(Y)} \cup 1 - R < \frac{f_X(a-Y)}{Mg_Y(Y)}\right]$$

$$= \frac{\displaystyle\int_0^x dy \int_0^{\frac{f_X(y)}{Mg_Y(y)}} f_{Y,R}(y,r)\,dr + \int_{a-x}^a dy \int_{1-\frac{f_X(a-y)}{g_Y(y)}}^1 f_{Y,R}(y,r)\,dr}{\displaystyle\int_0^a dy \int_0^{\frac{f_X(y)}{Mg_Y(y)}} f_{Y,R}(y,r)\,dr + \int_0^a dy \int_{1-\frac{f_X(a-y)}{g_Y(y)}}^1 f_{Y,R}(y,r)\,dr}.$$

Since

$$\int_0^{\frac{f_X(y)}{Mg_Y(y)}} f_{Y,R}(y,r)\,dr = g_Y(y) \int_0^{\frac{f_X(y)}{Mg_Y(y)}} dr = \frac{f_X(y)}{M}$$

and similarly for the other integrals, we have

$$P(X \leqslant x) = \frac{\displaystyle\int_0^x \frac{f_X(y)}{M}\,dy + \int_{a-x}^a \frac{f_X(a-y)}{M}\,dy}{\displaystyle\int_0^a \frac{f_X(y)}{M}\,dy + \int_0^a \frac{f_X(a-y)}{M}\,dy}$$

$$= \frac{\displaystyle\int_0^x \frac{f_X(y)}{M}\,dy + \int_0^x \frac{f_X(u)}{M}\,du}{\displaystyle\int_0^a \frac{f_X(y)}{M}\,dy + \int_0^a \frac{f_X(u)}{M}\,du}$$

$$= \frac{2F_X(x)/M}{2F_X(a)/M} \qquad (3.9)$$

or

$$P(X \leqslant x) = F_X(x).$$

We note from eqn (3.9) that the sampling efficiency is $2F_X(a)/M = 2/M$. For a uniform target distribution,

$$M \leqslant 2 \max [f_X(x)/g_Y(x)],$$

and so in this case the sampling efficiency can never be worse than that of the conventional envelope rejection method.

3.6   Ratio of uniforms method

This is a method due to Kinderman and Monahan (1977) which is based on the acceptance or rejection of a prospective variate, generated from the ratio of two random numbers. It relies on the following theorem.

THEOREM 3.1   Let $C = \{(u, v): 0 \leqslant u \leqslant f_X^{\frac{1}{2}}(v/u)\}$. Suppose points with coordinates (U, V) are uniformly distributed over C. Then the p.d.f. of V/U is $f_X(\cdot)$.

Proof.   Consider a transformation $(U, V) \rightarrow (U, Z)$ where $Z = V/U$. The Jacobian of this transformation is U. Thus, by the uniformity of (U, V) over C, the joint p.d.f. of U and Z is

$$f_{U,Z}(u, z) = \begin{cases} \dfrac{u}{\displaystyle\iint_C du\,dv} & (0 \leqslant u \leqslant f_X^{\frac{1}{2}}(z)) \\[2mm] 0 & \text{(elsewhere).} \end{cases}$$

Hence the marginal density of Z is

$$f_Z(z) = \frac{\displaystyle\int_0^{f_X^{\frac{1}{2}}(z)} u\,du}{\displaystyle\iint_C du\,dv} = \frac{\frac{1}{2}f_X(z)}{\displaystyle\iint_C du\,dv}.$$

Since $f_Z(\cdot)$ and $f_X(\cdot)$ are both p.d.f.'s, it follows that

$$\iint_C du\,dv = \tfrac{1}{2}$$

and

$$f_Z(z) = f_X(z),$$

showing that V/U has the desired distribution.

To generate from $f_X(\cdot)$, a conveniently shaped region D (usually, but not necessarily, rectangular or triangular) which completely encloses C is identified. A point (U, V) uniformly distributed over D is generated. If the point also lies within C, that is if $0 \leqslant U \leqslant f_X^{\frac{1}{2}}(V/U)$, the prospective variate $X = V/U$ is accepted, otherwise the prospective variate is rejected and a new attempt is made. Figure 3.3 shows a typical region C enclosed by a rectangle D.

The dimensions of a minimal enclosing rectangle for C may be obtained through the following lemma.

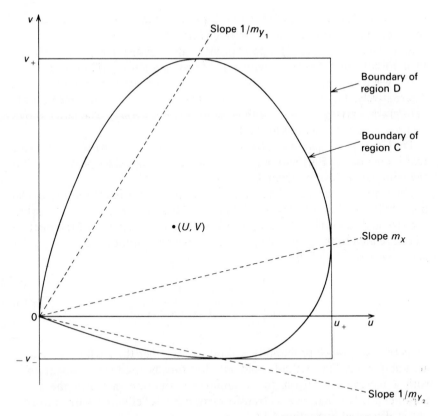

Fig. 3.3. Acceptance region C and (rectangular) enclosing region D

LEMMA A   Let X be a continuous random variable and $Y = X^{-1}$. Define m_X, m_{Y_1}, and m_{Y_2} by

$$f_X^{\frac{1}{2}}(m_X) = \max_x \{f_X^{\frac{1}{2}}(x)\}$$

$$f_Y^{\frac{1}{2}}(m_{Y_1}) = \max_{y \geqslant 0} \{f_Y^{\frac{1}{2}}(y)\}$$

$$f_Y^{\frac{1}{2}}(m_{Y_2}) = \max_{y \leqslant 0} \{f_Y^{\frac{1}{2}}(y)\}.$$

Let $u_+ = f_X^{\frac{1}{2}}(m_X)$, $v_+ = f_Y^{\frac{1}{2}}(m_{Y_1})$, and $v_- = f_Y^{\frac{1}{2}}(m_{Y_2})$. Then the minimal rectangle (with sides parallel to the axes) enclosing C is

$$D = \{(u, v): 0 \leqslant u \leqslant u_+; \ -v_- \leqslant v \leqslant v_+\}$$

Proof.   Suppose $(u, v) \in C$. This implies that $0 \leqslant u \leqslant f_X^{\frac{1}{2}}(v/u)$. From the definition of m_X and u_+, this implies that $0 \leqslant u \leqslant u_+$. Since $u = u_+$ when $v/u = m_X$ the line $u = u_+$ is tangential to the boundary of C.

Since $f_Y(y) = y^{-2} f_X(y^{-1})$, a point $(u, v) \in C$ also satisfies $|v| \leqslant f_Y^{\frac{1}{2}}(u/v)$. If $v \geqslant 0$, then $u/v \geqslant 0$ and $f_Y^{\frac{1}{2}}(u/v) \leqslant v_+$ with equality when $u/v = m_{Y_1}$. Thus $v \leqslant v_+$ with $v = v_+$ when $u/v = m_{Y_1}$, showing that $v = v_+$ is tangential to the boundary of C. Alternatively, if $v < 0$, $|v| \leqslant f_Y^{\frac{1}{2}}(u/v)$ implies that $v \geqslant -f_Y^{\frac{1}{2}}(u/v)$. But since $u/v \leqslant 0$, $f_Y^{\frac{1}{2}}(u/v) \leqslant v_-$, giving $v \geqslant -v_-$ with equality when $u/v = m_{Y_2}$. The latter shows that $v = -v_-$ is tangential to the boundary of C.

Thus any point $(u, v) \in C$ satisfies $0 \leqslant u \leqslant u_+$ and $-v_- \leqslant v \leqslant v_+$, showing that D encloses C. Further the tangency of the rectangle sides proves that it is the minimal enclosing rectangle.

The lemma shows that the rectangle side lengths are related to the modal probability of X and to two modal probabilities of Y, one for non-negative variates and the other for non-positive values. The efficiency of the method will be related to the probability that a prospective variate is accepted, and is given by

$$P = \frac{\iint_C du\, dv}{\iint_D du\, dv} = \frac{\frac{1}{2}}{u_+(v_+ + v_-)}. \tag{3.10}$$

Thus the acceptance probability will be zero if either the p.d.f. of X or X^{-1} are unbounded. The method could not, therefore, be used in its present form, with a rectangular region, for a gamma distribution in which the shape parameter is less than one. A transformation $Z = X^\alpha$ eliminates this problem and is discussed in Section 4.3.6.

To illustrate use of the ratio method, we consider generation from a (standardized) gamma distribution

$$f_X(x) = \frac{x^{\alpha - 1} e^{-x}}{\Gamma(\alpha)} \qquad (x \geqslant 0),$$

where the shape parameter $\alpha \geqslant 1$. The mode m_X lies at $\alpha - 1$, while the density of the reciprocal variate $Y = X^{-1}$ is

$$f_Y(y) = \frac{y^{-(\alpha + 1)} e^{-1/y}}{\Gamma(\alpha)} \qquad (y \geqslant 0).$$

Since Y cannot take negative values, $m_{Y_2} = 0$. Maximization of $f_Y(y)$ shows that $m_{Y_1} = (\alpha + 1)^{-1}$. Thus the rectangle dimensions are

$$u_+ = (\alpha - 1)^{\{(\alpha - 1)/2\}} e^{-\{(\alpha - 1)/2\}} / \{\Gamma(\alpha)\}^{\frac{1}{2}},$$

$$v_+ = (\alpha + 1)^{\{(\alpha + 1)/2\}} e^{-\{(\alpha + 1)/2\}} / \{\Gamma(\alpha)\}^{\frac{1}{2}},$$

$$v_- = 0.$$

Given two random numbers R_1 and R_2, points (U, V) uniformly distributed in C are obtained by setting $U = u_+ R_1$ and $V = v_+ R_2$. The prospective variate

$$X = \frac{V}{U} = \frac{R_2}{R_1} \cdot \frac{(\alpha+1)^{\{(\alpha+1)/2\}}}{(\alpha-1)^{\{(\alpha-1)/2\}}e}$$

and the acceptance condition $0 \leqslant U \leqslant f_X^{\frac{1}{2}}(V/U)$ reduces to:

$$R_1 u_+ \leqslant f_X^{\frac{1}{2}}(X)$$

or

$$R_1 \leqslant \left(\frac{X}{\alpha-1}\right)^{\{(\alpha-1)/2\}} e^{-\frac{1}{2}(X-\alpha+1)}.$$

As in envelope rejection, the p.d.f. $f_X(\cdot)$ is required to be known only up to a multiplicative constant. The acceptance probability may be obtained from eqn (3.10) giving

$$P = \frac{e^\alpha \Gamma(\alpha)}{2(\alpha^2-1)^{(\alpha/2)}} \left(\frac{\alpha-1}{\alpha+1}\right)^{\frac{1}{2}}.$$

Figure 3.4 shows C and D for various values of α. As α increases, the acceptance region C occupies a smaller proportion of the enclosing region D. When $\alpha = 1$, $P = e/4$. For large α, we may use Stirling's approximation,

$$\Gamma(\alpha) = \sqrt{(2\pi)} e^{-(\alpha-1)}(\alpha-1)^{\alpha-\frac{1}{2}},$$

to show that $P \simeq \sqrt{\{\pi/(2\alpha)\}}$. This behaviour is confirmed by the diagrams for large α, where the acceptance region is 'cigar-shaped' and is concentrated around the region $m_X \leqslant v/u \leqslant m_{Y_1}^{-1}$ or $\alpha - 1 \leqslant v/u \leqslant \alpha + 1$. This problem of low efficiency is likely to occur whenever we have a distribution with small coefficient of variation.

One way to improve the acceptance probability is to select a tighter fitting enclosing region. A possible choice is a parallelogram as shown for $\alpha = 20$ in Fig. 3.4. Such a method has been used as the basis of an efficient generator (Cheng and Feast, 1979). An alternative method is to force an increase in the coefficient of variation. This has been suggested for the gamma distribution by Kinderman and Monahan (1980), who generate relocated variates $X' = X - \alpha + 1$. The p.d.f. of X' and the associated reciprocal variate $Y' = 1/X'$ are thus

$$f_{X'}(x) = \frac{(x+\alpha-1)^{\alpha-1} e^{-(x+\alpha-1)}}{\Gamma(\alpha)} \qquad (x \geqslant 1-\alpha)$$

and

$$f_{Y'}(y) = \frac{(y^{-1}+\alpha-1)^{\alpha-1} e^{-(y^{-1}+\alpha-1)} y^{-2}}{\Gamma(\alpha)} \qquad (y \geqslant 0, y \leqslant (1-\alpha)^{-1}).$$

We may obtain the enclosing rectangle dimensions using Lemma A, as

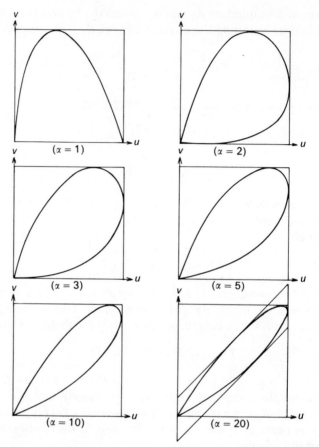

Fig. 3.4. Acceptance region C and enclosing region D for gamma, ratio of uniforms method

before. Maximizing $f_{X'}(x)$ and $f_{Y'}(y)$ (subject to $y \geqslant 0$ and $y \leqslant 0$) respectively gives

$$m_{X'} = 0 \qquad \text{(mode relocated to origin)},$$

$$m_{Y'_1} = \{1 + \sqrt{(2\alpha - 1)}\}^{-1},$$

$$m_{Y'_2} = \{1 - \sqrt{(2\alpha - 1)}\}^{-1}, \tag{3.11}$$

$$u_+ = (\alpha - 1)^{\{(\alpha - 1)/2\}} e^{-(\alpha - 1)/2} / \{\Gamma(\alpha)\}^{\frac{1}{2}},$$

$$v_+ = \frac{\{\alpha + \sqrt{(2\alpha - 1)}\}^{(\alpha - 1)/2} e^{-\{\alpha + \sqrt{(2\alpha - 1)}\}/2} \{\sqrt{(2\alpha - 1)} + 1\}}{\{\Gamma(\alpha)\}^{\frac{1}{2}}},$$

$$v_- = \frac{\{\alpha - \sqrt{(2\alpha - 1)}\}^{(\alpha - 1)/2} e^{-\{\alpha - \sqrt{(2\alpha - 1)}\}/2} \{\sqrt{(2\alpha - 1)} - 1\}}{\{\Gamma(\alpha)\}^{\frac{1}{2}}}.$$

Points within D are obtained by setting $U = u_+ R_1$ and $V = R_2(v_+ + v_-) - v_-$.
Using (3.11) a prospective relocated variate $X' = V/U$ is given by

$$X' = \frac{aR_2}{R_1} + \frac{b(1 - R_2)}{R_1}$$

where

$$a = \{1 + \sqrt{(2\alpha - 1)}\} \left\{ \frac{\alpha + \sqrt{(2\alpha - 1)}}{\alpha - 1} \right\}^{(\alpha - 1)/2} e^{-\{1 + \sqrt{(2\alpha - 1)}\}/2}$$

$$b = 2(1 - \alpha)/(ea).$$

The original gamma variate is obtained by setting $X = X' + \alpha - 1$. Clearly $X < 0$ is a sufficient condition for rejection. Otherwise the acceptance condition is $U \leq f_{X'}^{\frac{1}{2}}(X')$, that is

$$u_+ R_1 \leq f_{X'}^{\frac{1}{2}}(X')$$

or

$$R_1 \leq \left(\frac{X}{\alpha - 1} \right)^{(\alpha - 1)/2} e^{-X'/2}. \tag{3.12}$$

Figure 3.5 shows the acceptance regions for various α. Relocation of the mode to zero results in a high acceptance probability for all α. When $\alpha = 1$, it is $e/4$ as before (since no relocation is involved). For large α, consider the acceptance probability for a variate $X'/\sqrt{\alpha}$, which is asymptotically standard Normal. Lemma A can be used to show that this is $\sqrt{(\pi e)}/4$.

The illustrations given use *rectangular* enclosing regions. Higher acceptance probabilities for ratio methods can sometimes be obtained by devising a tighter fitting enclosing region, such as a polygon. In this respect an efficient method due to Hsuan (1979) for generating points uniformly within a polygon without rejection may prove useful.

How good is the ratio of uniforms procedure as a general variate generation method? The set-up time is unlikely to be much greater than a few marginal generation times. Regarding marginal generation time, an attractive feature is that the prospective variate is obtained through a ratio of two random numbers, and does not involve more time-consuming operations. The overall marginal generation time will depend upon the acceptance probability P and the number of time-consuming function evaluations in the acceptance test. The latter can often be reduced by judicious use of pre-tests as in the 'squeeze' technique. There is obviously a trade-off here between generation time and program complexity. There are no large memory requirements. Providing the mathematics is reduced to its simplest form and random numbers in the open (0, 1) rather than the closed [0, 1) interval are used, there should be no numerical problems to compromise statistical reliability, and typically programs are short, enabling them to be coded quickly and accurately.

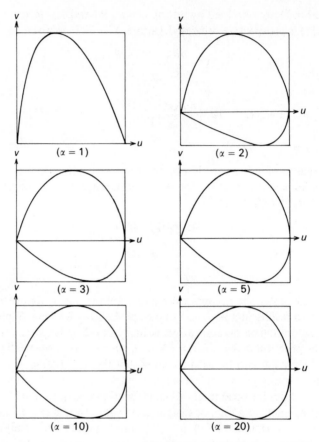

Fig. 3.5. Acceptance region C and enclosing region D for gamma, relocated ratio of uniforms method

3.7   Comparison of random numbers (Forsythe's method)

This method is based on one alluded to by von Neumann in 1949 in a procedure for generating negative exponential variates, using just a comparison of random numbers. At the time, von Neumann suggested that the method could be generalized to generate from any p.d.f. satisfying a first-order differential equation. Forsythe (1972) gives an exposition of what he assumes von Neumann had in mind. The technique is usually referred to as *Forsythe's method*. We consider first the following theorem.

THEOREM 3.2   Let X_j be a random variable uniformly distributed in the interval (q_{j-1}, q_j) and $h_j(x)$ be a known continuous-valued function of x with $0 \leqslant h_j(x) \leqslant 1$ within the interval. Let $\{R_i\}$ denote a sequence of random numbers, and N be an integer such that:

$$N = 1 \text{ if } h_j(X_j) < R_1,$$

$$N = n(>1) \text{ if } h_j(X_j) \geqslant R_1 \ldots \geqslant R_{n-1} < R_n.$$

Then the p.d.f. of X_j, conditional upon N being odd-valued is

$$\psi_j(x) = \frac{e^{-h_j(x)}}{\displaystyle\int_{q_{j-1}}^{q_j} e^{-h_j(u)} \, du} \qquad (q_{j-1} \leqslant x \leqslant q_j).$$

Proof.   Let $g_{X_j, N}(x, n)$ be the joint probability mass/density function of X_j and N, where it is understood that the function is zero outside the interval (q_{j-1}, q_j). Then

$$g_{X_j, N}(x, n) = \begin{cases} (q_j - q_{j-1})^{-1} P(h_j(X_j) \geqslant R_1 \ldots \geqslant R_{n-1} < R_n | X_j = x) & (n > 1) \\ (q_j - q_{j-1})^{-1} P(h_j(X_j) < R_1 | X_j = x) & (n = 1) \end{cases}$$

$$= \begin{cases} (q_j - q_{j-1})^{-1} \displaystyle\int_0^{h_j(x)} dr_1 \int_0^{r_1} dr_2 \ldots \int_0^{r_{n-2}} dr_{n-1} \int_{r_{n-1}}^1 dr_n & (n > 1) \\ (q_j - q_{j-1})^{-1} \{1 - h_j(x)\} & (n = 1) \end{cases}$$

or

$$g_{X_j, N}(x, n) = (q_j - q_{j-1})^{-1} \left[\frac{\{h_j(x)\}^{n-1}}{(n-1)!} - \frac{\{h_j(x)\}^n}{n!} \right] \qquad (n \geqslant 1). \qquad (3.13)$$

The density of X_j, given that N is odd-valued is

$$\psi_j(x) = \frac{\displaystyle\sum_{n \text{ odd}} g_{X_j, N}(x, n)}{\displaystyle\int_{q_{j-1}}^{q_j} dx \sum_{n \text{ odd}} g_{X_j, N}(x, n)}.$$

From eqn (3.13)

$$\sum_{n \text{ odd}} g_{X_j, N}(x, n) = (q_j - q_{j-1})^{-1} e^{-h_j(x)} \qquad (3.14)$$

giving

$$\psi_j(x) = e^{-h_j(x)} \Bigg/ \int_{q_{j-1}}^{q_j} e^{-h_j(u)} \, du,$$

which completes this proof.

Suppose now that we wish to sample from a p.d.f. $f_X(\cdot)$. This may be expressed as a probability mixture,

$$f_X(x) = \sum_j p_j \psi_j(x) \tag{3.15}$$

where

$$\psi_j(x) = \begin{cases} e^{-h_j(x)} \Big/ \int_{q_{j-1}}^{q_j} e^{-h_j(u)} du & (q_{j-1} \leqslant x \leqslant q_j) \\ 0 & \text{(elsewhere)}, \end{cases} \tag{3.16}$$

$$p_j = \int_{q_{j-1}}^{q_j} f_X(u) du,$$

and $q_0 < q_1 < q_2 \ldots < q_{j-1} < q_j < \ldots$ The critical requirement for (3.16) to be a useful representation in this instance is that the intervals (q_{j-1}, q_j) and the functions $h_j(x)$ be chosen in such a way that

$$0 \leqslant h_j(x) \leqslant 1 \qquad (q_{j-1} \leqslant x \leqslant q_j). \tag{3.17}$$

Given the representation (3.15), the sampling procedure is to use the composition method, generating from $\psi_j(x)$ with probability p_j. Given a random number R, the correct interval (q_{j-1}, q_j) is selected by finding the smallest j such that

$$\sum_{s=1}^{j} p_s \geqslant R.$$

Sampling from $\psi_j(x)$ is accomplished by accepting X_j if N is odd, otherwise rejecting it.

The advantage of Forsythe's method is that since $h_j(x) = -\ln[K\psi_j(x)]$ where K is a constant, it is often very quick to evaluate. This is the case for example in the normal, exponential, and von Mises distributions. Further, the comparisons in Theorem 3.2 can be executed rapidly, providing N is not too large. A disadvantage of the method is that in most cases a table of p_j values must be calculated (often using numerical integration) and stored. An exception to this is the negative exponential distribution, which is discussed in Section 4.1. One major advantage of Forsythe's method is that by reserving one random number stream for interval selection, antithetic variates (see Section 3.1) may be generated.

In assessing the efficiency of any algorithm using this method, it is useful to compute the expected number of random numbers required to generate one variate from $\psi_j(x)$ as follows. The number of random numbers associated with each prospective variate is $N+1$, since N comparisons are needed and 1 random number required to generate X_j. The number of prospective variates required to obtain one (accepted) variate is Geometrically distributed with mean $1/P(N \text{ odd})$. Thus the mean number of random numbers required per accepted variate is $M_j = E(N+1)/P(N \text{ odd})$. From eqn (3.14),

$$P(N \text{ odd}) = (q_j - q_{j-1})^{-1} \int_{q_{j-1}}^{q_j} e^{-h_j(x)} dx.$$

From eqn (3.13),

$$E(N+1) = (q_j - q_{j-1})^{-1} \int_{q_{j-1}}^{q_j} \sum_1^\infty (n+1) \left[\frac{\{h_j(x)\}^{n-1}}{(n-1)!} - \frac{\{h_j(x)\}^n}{n!} \right] dx$$

$$= (q_j - q_{j-1})^{-1} \int_{q_{j-1}}^{q_j} (e^{h_j(x)} + 1) dx.$$

Hence,

$$M_j = \frac{(q_j - q_{j-1}) + \int_{q_{j-1}}^{q_j} e^{h_j(x)} dx}{\int_{q_{j-1}}^{q_j} e^{-h_j(x)} dx}.$$

When implementing the method, it is worthwhile to use the same random number to select both the interval j and the first prospective variate within the interval. (The latter may be accomplished by utilizing the conditional uniformity of the random number between $\sum_{s=1}^{j} p_s$ and $\sum_{s=1}^{j+1} p_s$). If this is done, the mean number of random numbers required to generate one variate from $f_X(\cdot)$ is

$$M = \sum_j p_j M_j.$$

M can always be made arbitrarily close to two, by selecting narrow enough intervals. In such cases $h_j(x) \simeq 0$ throughout the interval, giving $N = 1$ on most occasions. There is a trade-off here between the amount of work done in each interval and the number of intervals to be set up and searched through.

To illustrate the method, consider generation from a folded standard normal distribution with p.d.f.

$$f_X(x) = \left(\frac{2}{\pi} \right)^{\frac{1}{2}} e^{-\frac{1}{2}x^2} \qquad (x \geq 0).$$

Generation from the usual standard normal distribution follows by assigning a random sign to X. Using the representation (3.15) we define

$$h_j(x) = \tfrac{1}{2}(x^2 - q_{j-1}^2) \qquad (q_{j-1} \leq x \leq q_j).$$

If a high priority attaches to using few intervals, then the aim should be to maximize interval widths (subject to $h_j(x) \leqslant 1$). Since $h_j(x)$ is an increasing function of x, this is equivalent to solving the difference equations:

$$q_0 = 0$$

$$\tfrac{1}{2}(q_j^2 - q_{j-1}^2) = 1. \tag{3.18}$$

The solution to these is

$$q_j = (2j)^{\frac{1}{2}}.$$

Hence

$$p_j = \int_{\{2(j-1)\}^{\frac{1}{2}}}^{(2j)^{\frac{1}{2}}} \left(\frac{2}{\pi}\right)^{\frac{1}{2}} e^{-\frac{1}{2}x^2} dx \qquad (j = 1, 2, \ldots, k).$$

The number of intervals k, and therefore the truncation point for the distribution, may be chosen so that the tail probability is negligible. The alternative is to select a tail probability which is not necessarily negligible and to use an exact distribution tail generation method as described in Section 7.3. For example, if the tail probability is not to exceed 2×10^{-4}, then from standard normal tables we require the smallest k such that $3.72 \leqslant \sqrt{(2k)}$, that is seven intervals.

Forsythe (1972) and Atkinson and Pearce (1976) both use a different set of intervals, defined by

$$q_0 = 0,$$

$$q_j = (2j-1)^{\frac{1}{2}} \qquad (j \geqslant 1). \tag{3.19}$$

They claim that these give maximum width. Clearly this is not the case, since the first interval has width 1, compared with $\sqrt{2}$ under (3.18). This error appears to stem from a misleading statement by Forsythe that the interval widths should not exceed one. In fact, providing $h_j(x) \leqslant 1$ within an interval there is no upper bound on interval widths. Using the intervals defined by eqn (3.19), a tail probability not exceeding 2×10^{-4} would require the smallest k satisfying

$$3.72 \leqslant (2k-1)^{\frac{1}{2}},$$

a requirement of eight intervals compared with seven, under (3.18). This of course does not necessarily favour the maximum width scheme, particularly as narrow intervals require fewer random numbers per interval. Offset against this is the fact that a larger number of intervals will require slightly more time to select the appropriate interval.

In summary, Forsythe's method is capable of producing very fast routines (in the limit two random numbers only are required) at the expense of potentially high set-up times and storage of a large number of constants. The routines will frequently involve numerical integrations, making them less portable. In these cases the approach is unlikely to be used unless marginal generation time is of extreme importance and the distribution parameters remain fixed between calls. If the p.d.f. or its domain are unbounded, then an infinite number of intervals is required. In practice this will mean (exponential excluded) some truncation of the distribution, which may affect the generator's statistical reliability. Finally the method has not been applied in any direct way to discrete distributions.

3.8 Alias rejection method

The alias rejection method devised by Walker (1977) is an efficient procedure for generating discrete random variables having a finite number (m) of mass points. It is in part an envelope rejection procedure, with first-stage sampling from a uniform distribution. However, if the acceptance test is not passed, the prospective variate undergoes a deterministic transformation, the transformed variate being accepted. It is therefore a highly attractive method in that 'rejected values' are not lost.

Kronmal and Peterson (1979) give a theoretical justification for the procedure which does not appear in Walker's account. They show that any m-point discrete distribution may be represented by a certain equi-probable mixture of m two-point distributions.

To demonstrate this, consider, without loss of generality, a discrete distribution $f_m(x)$ ($x = 0, 1, 2, \ldots, m-1$). Then $f_m(x)$ may be represented as the following mixture of a $(m-1)$-point distribution $f_{m-1}(x)$ and a two-point distribution $g_{j_m}(x)$:

$$f_m(x) = \left(\frac{m-1}{m}\right) f_{m-1}(x) + \left(\frac{1}{m}\right) g_{j_m}(x) \tag{3.19}$$

where,

$$f_{m-1}(x) = \begin{cases} \left(\dfrac{m}{m-1}\right) f_m(x) & (x \neq j_m, a_{j_m}) \\ 0 & (x = j_m) \\ \left(\dfrac{m}{m-1}\right) [f_m(j_m) + f_m(a_{j_m})] - \left(\dfrac{1}{m-1}\right) & (x = a_{j_m}) \end{cases} \tag{3.20}$$

and

$$g_{j_m}(x) = \begin{cases} 0 & (x \neq j_m, a_{j_m}) \\ m f_m(j_m) & (x = j_m) \\ 1 - m f_m(j_m) & (x = a_{j_m}), \end{cases} \tag{3.21}$$

j_m and a_{j_m} being values belonging to the domain of $f_m(x)$ such that

$$f_m(j_m) \leqslant \left(\frac{1}{m}\right) \tag{3.22}$$

and

$$f_m(a_{j_m}) \geqslant \left(\frac{1}{m}\right). \tag{3.23}$$

To establish the validity of the representation (3.19)–(3.23) we need demonstrate only: (i) that a pair j_m and a_{j_m} satisfying (3.22) and (3.23) will always exist; (ii) that $f_{m-1}(x)$ is a bona fide probability distribution; (iii) that $g_{j_m}(x)$ is a bona fide probability distribution. (i) follows otherwise $\sum_x f_m(x) \neq 1$; (ii) follows since $f_{m-1}(x) \geqslant 0$ $(x \neq a_{j_m})$, $f_{m-1}(a_{j_m}) \geqslant 0$ as a consequence of (3.23) and $\sum_x f_{m-1}(x) = 1$; (iii) follows since $g_{j_m}(j_m) \geqslant 0$, $g_{j_m}(a_{j_m}) \geqslant 0$ as a consequence of (3.22), and $\sum_x g_{j_m}(x) = 1$.

Applying the representation (3.19)–(3.23) recursively $(m-1)$ times we find that

$$f_m(x) = \left(\frac{1}{m}\right) \sum_{j=0}^{m-1} g_j(x) \tag{3.24}$$

where

$$g_j(x) = \begin{cases} 0 & (x \neq j, a_j) \\ c_j & (x = j) \\ 1 - c_j & (x = a_j). \end{cases} \tag{3.25}$$

Thus $f_m(x)$ is an equi-probable mixture of m two-point distributions. Note that one of the mass points of $g_j(x)$ is j itself, with probability c_j (the 'cut-off' value). The other mass point is a_j, the 'alias' of j. The cut-off and alias values are determined algorithmically through eqns (3.19)–(3.23). Kronmal and Peterson (1978) give a 40-statement FORTRAN listing for such a procedure, and note that execution time is proportional to m, the number of mass points.

Once the setting up of c_j and a_j values is effected, the sampling routine takes the simple form shown in Algorithm 3.5.

Algorithm 3.5

Input $[m, a_0, \ldots, a_{m-1}, c_0, \ldots, c_{m-1}]$
Output $[X]$

1 generate $R \sim U(0, 1)$.
2 $X := Rm$.
3 If $X - \langle X \rangle \leqslant c_{\langle X \rangle}$ then
4      $X := \langle X \rangle$
5 Else
6      $X := a_{\langle X \rangle}$
7 End If.
8 Exit.

Note that $\langle X \rangle$ is uniformly distributed on the integers $[0, m-1]$, that $X - \langle X \rangle \sim U(0,1)$, and that consequently step 3 ensures that $\langle X \rangle$ is delivered with probability $c_{\langle X \rangle}$, and $a_{\langle X \rangle}$ with probability $1 - c_{\langle X \rangle}$, as required. Only one random number is required per variate generation. The method as described by Walker and Kronmal and Peterson can be applied only to distributions having a finite number of mass points. For distributions having infinite domain the p.m.f. can be represented as a probability mixture of two mass functions, one having finite domain (where the alias method can be applied) and the other covering the tail of the distribution (see Section 5.4).

An analogous procedure to the alias rejection method for continuous distributions is the acceptance complement method (Kronmal and Peterson, 1981). Suppose it is required to generate variates X where

$$f_X(x) = p f_Z(x) + (1-p) f_W(x) \qquad (0 \leqslant p \leqslant 1).$$

Let $h_U(\cdot)$ be any p.d.f. dominating $p f_Z(\cdot)$. Then Algorithm 3.6 generates X-values. It is clear that accepting W, following rejection of U, is analogous to acceptance of the alias value in the discrete case. To establish the validity of the algorithm, we note that the joint p.d.f. of R, U, and W is $h_U(\cdot) f_W(\cdot)$. Hence

$$P(X \leqslant x) = \int_{-\infty}^{\infty} f_W(w)\, dw \int_{-\infty}^{x} h_U(u)\, du \int_0^{p f_Z(u)/h_U(u)} dr$$
$$+ \int_{-\infty}^{x} f_W(w)\, dw \int_{-\infty}^{\infty} h_U(u)\, du \int_{p f_Z(u)/h_U(u)}^{1} dr.$$

Since $h_U(u) \geqslant p f_Z(u)$,

$$\int_0^{p f_Z(u)/h_U(u)} dr = p f_Z(u)/h_U(u)$$

Algorithm 3.6

Input $[f_Z(\cdot), h_U(\cdot), f_W(\cdot), p]$
Output $[X]$

1   generate $U \sim h_U(\cdot)$ and $R \sim U(0,1)$.
2   If $R < p f_Z(U)/h_U(U)$ then
3       $X := U$
4   Else
5   generate $W \sim f_W(\cdot)$
6       $X := W$
7   End If.
8   Exit.

and hence

$$P(X \leqslant x) = \int_{-\infty}^{x} p f_Z(u)\, du + \int_{-\infty}^{\infty} [h_U(u) - p f_Z(u)]\, du \int_{-\infty}^{x} f_W(w)\, dw,$$

whence

$$f_X(x) = p f_Z(x) + (1-p) f_W(x),$$

as required.

The method can be thought of as a variant of envelope rejection. The flexibility afforded by choice of densities for U and Z would seem to be an advantage. However, examples where it outperforms a good envelope procedure, both from a time and ease of implementation viewpoint, are hard to find.

3.9 Polynomial sampling

This idea has been used in connection with the generation of normal variates (Ahrens and Dieter, 1972), but would appear to be of wider applicability. Suppose we wish to generate variates having p.d.f.

$$f_X(x) = \sum_{j=0}^{k} b_j x^j \qquad (0 \leqslant x \leqslant 1), \tag{3.26}$$

where the coefficients $\{b_j\}$ are not necessarily all non-negative. Define sets S and T as,

$$S = \{j \mid b_j \geqslant 0; j \neq 0\}; \ T = \{j \mid b_j < 0; j \neq 0\}.$$

Then eqn (3.26) may be expressed as

$$f_X(x) = \sum_{j=0}^{k} w_j f_{X_j}(x), \tag{3.27}$$

where

$$f_{X_0}(x) = 1 \tag{3.28}$$

$$f_{X_j}(x) = \begin{cases} (j+1) x^j & (j \in S) \\ \left(\dfrac{j+1}{j}\right)(1-x^j) & (j \in T) \end{cases} \tag{3.29}$$

$$w_0 = b_0 + \sum_{j \in T} b_j,$$

and

$$w_j = \begin{cases} b_j/(j+1) & (j \in S) \\ -j b_j/(j+1) & (j \in T). \end{cases} \tag{3.30}$$

We note that $\{f_{X_j}(\cdot)\}$ are bona fide p.d.f.s, that $w_j \geqslant 0$ ($j \geqslant 1$), and that $w_0 \geqslant 0$ providing

$$b_0 + \sum_{j \in T} b_j \geqslant 0. \tag{3.31}$$

Result (3.31) is the condition for eqn (3.27) to be a valid probability mixture representation of (3.26). If (3.31) is satisfied, then sampling can proceed by generating X_j with probability w_j. Unless all the $\{b_j\}$ are non-negative, then sampling from the $\{f_{X_j}(\cdot)\}$ is non-trivial, since some of the associated c.d.f.s cannot be inverted analytically.

However, the following method involving the generation of $(j+1)$ random numbers can be used to sample from $f_{X_j}(\cdot)$, for both $j \in S$ and $j \in T$. Let $\{R_1, R_2, \ldots, R_{j+1}\}$ be a set of $(j+1)$ random numbers and

$$V \sim U(0, U),$$

where

$$U = \max(R_1, R_2, \ldots, R_{j+1}).$$

Then the c.d.f.s of U and V are

$$F_U(x) = x^{j+1} \qquad (0 \leqslant x \leqslant 1),$$

and

$$F_V(x) = \int_x^1 (x/u)(j+1)u^j \, du + \int_0^x (j+1)u^j \, du$$

$$= \left(\frac{j+1}{j}\right) x - \frac{x^{j+1}}{j} \qquad (0 \leqslant x \leqslant 1),$$

which are identical to the c.d.f.s associated with eqn (3.29).

Algorithm 3.7 generates variates from p.d.f.s (3.28) and (3.29). If $j = 0$, then X is simply delivered as a random number. If $j \neq 0$ and $b_j \geqslant 0$, then U, the maximum of $(j+1)$ random numbers is delivered. If $j \neq 0$ and $b_j < 0$, then V is generated by delivering R_{j+1}, if it is not the maximum of $R_1, R_2, \ldots, R_j$, R_{j+1}, otherwise R_j, the previous random number is delivered.

Although Algorithm 3.7 is readily implemented, it is perhaps the calculation of $\{w_j\}$ values which has detracted from the general use of polynomial sampling in the past. The statistical reliability will depend on the extent to which a finite degree polynomial can approximate $f_X(\cdot)$. k may have to be large in order to provide a good fit. Since the $\{w_j\}$ are obtained from coefficients in a Maclaurin expansion of $f_X(\cdot)$, this may involve considerable set-up time and the storage of a large number of constants. Another possible problem is that the domain of $f_X(\cdot)$ is too large to support the necessary condition, $w_0 \geqslant 0$. One way round this is to split the domain into separate intervals. There is an analogy here with Forsythe's method in which the same trick may be needed to force $h_j(x) \leqslant 1$. If these problems are surmountable,

Algorithm 3.7

Input $[j, b_j]$
Output $[X]$

1   If $j=0$ then
2   generate $R \sim U(0, 1)$
3        $X := R$
4   Exit
5   End If.
6    $R := 0$, $Rmax := -\infty$.
7   For $k := 1, j+1$
8        $Rprev := R$
9   generate $R \sim U(0, 1)$
10   If $R > Rmax$ then
11           $Rmax := R$
12   End If.
13   Next k.
14   $Rfinal := R$.
15   If $b_j \geqslant 0$ then
16       $X := Rmax$
17   Else If $Rfinal < Rmax$ then
18       $X := Rfinal$
19   Else
20       $X := Rprev$
21   End If.
22   Exit.

the performance of the generator is governed by its marginal generation time. Since only comparisons are involved, this will depend mainly on $1 + \sum_{j=0}^{k} (j+1)w_j$, the mean number of random numbers required per variate. Using eqns (3.26) and (3.30) this becomes

$$1 + f_X(1) - \sum_{j \in T} jb_j$$

which reduces to $1 + f_X(1)$, in the simple case when the polynomial has no negative coefficients.

3.10 Transformations involving multiple roots

Suppose it is convenient to generate variates Y from some specified distribution, but that the main interest is in generating X values where $Y = g(X)$ is a known function. If the inverse $g^{-1}(\cdot)$ is single-valued everywhere, then in principle there should be no problem. Difficulties arise if a value $Y=y$ gives rise to an equation $g(x)=y$, which has multiple roots. We will suppose that there are m distinct roots $x_1(y), \ldots, x_m(y)$, where m may depend upon y. The problem is to determine which of the roots shall be delivered as the variate X.

Michael et al. (1976) considered both discrete and continuous cases. Let X and Y be discrete random variables. Then if $Y=y$, the root $x_i(y)$ should be returned with (conditional) probability,

$$
\begin{aligned}
p_i(y) &= P(X=x_i(y)|g(X)=y) \\
&= \frac{P(X=x_i(y), g(X)=y)}{P(g(X)=y)} \\
&= \frac{P(X=x_i(y))}{\sum_{j=1}^{m} P(X=x_j(y))} \\
&= \left\{1+\sum_{j\neq i} \frac{P(X=x_j(y))}{P(X=x_i(y))}\right\}^{-1}.
\end{aligned}
\tag{3.32}
$$

If X and Y are continuous random variables and $g(\cdot)$ is a suitably well-behaved function, then

$$
p_i(y)= \lim_{h\to 0+} \{P(x_i-\Delta_i(h)\leqslant X<x_i+\Delta_i(h)|y-h\leqslant g(X)<y+h)\} \tag{3.33}
$$

where $[x_i-\Delta_i(h), x_i+\Delta_i(h))$ is the inverse image of $[y-h, y+h)$, which contains the ith root of $g(x)=y$. Since $\Delta_i(h)=h/|g'(x_i)|+o(h)$, (3.33) becomes

$$
\begin{aligned}
p_i(y) &= \lim_{h\to 0+} \left\{\frac{P(x_i-\Delta_i(h)\leqslant X<x_i+\Delta_i(h))}{P(y-h\leqslant g(x)<y+h)}\right\} \\
&= \frac{f_X(x_i)/|g'(x_i)|}{\sum_{j=1}^{m} f_X(x_j)/|g'(x_j)|} \\
&= \left\{1+\sum_{j\neq i} \frac{f_X(x_j)}{f_X(x_i)}\left|\frac{g'(x_i)}{g'(x_j)}\right|\right\}^{-1}.
\end{aligned}
\tag{3.34}
$$

To generate a variate X it is necessary only to compute $\{p_i(y)\}$ and to return the ith root with probability $p_i(y)$. The main applications appear to be for continuous variates, and we shall illustrate with reference to Cauchy and standardized Wald distributions.

3.10.1 Cauchy distribution

Suppose we wish to generate X-values from a folded Cauchy distribution,

$$
f_X(x)=\frac{2}{\pi(1+x^2)} \qquad (x\geqslant 0). \tag{3.35}
$$

It will be convenient to take

$$Y = g(X) = \min(X, X^{-1}), \qquad (3.36)$$

giving

$$F_Y(y) = \begin{cases} 1 - P(y \leqslant X \leqslant y^{-1}) & (0 \leqslant y < 1) \\ 1 & (y \geqslant 1) \end{cases}$$

and hence

$$f_Y(y) = f_X(y) + y^{-2} f_X(y^{-1}) \qquad (0 \leqslant y < 1). \qquad (3.37)$$

Note that the transformed p.d.f. has *finite domain*, which in itself often makes the sampling process easier. From eqns (3.35) and (3.37)

$$f_Y(y) = \frac{4}{\pi(1 + y^2)} \qquad (0 \leqslant y < 1).$$

Given that $Y = y$, the two roots of (3.36) are

$$x_1 = y, \; x_2 = y^{-1}$$

giving

$$g'_X(x_1) = 1 \quad \text{and} \quad g'_X(x_2) = -x_2^{-2} = -y^2.$$

Further

$$\frac{f_X(x_2)}{f_X(x_1)} = \frac{1 + x_1^2}{1 + x_2^2} = y^2.$$

Using (3.34),

$$p_i(y) = 0.5 \qquad (i = 1, 2),$$

which in this case happens to be independent of y. The only remaining task is to decide how to sample from $f_Y(\cdot)$. It would be quite in order to invert the c.d.f. Since this involves a tangent evaluation, it offers no advantage over the standard inversion method for a Cauchy distribution. However the finite domain of $f_Y(\cdot)$ suggests use of envelope rejection with a uniform target distribution, and in this case gives an acceptance probability of $\pi/4$. Algorithm 3.8 gives the details. Steps 1 and 2 produce a Y-variate, while steps 3–7 deliver the correct root. A mean of $(8/\pi) + 1$ random numbers is required per variate, and all other operations are elementary. We may therefore expect the method to be faster than conventional inversion if the time to generate three random numbers is less than the time to evaluate a tangent.

Algorithm 3.8

Output [X]

1 generate $R_1, R_2 \sim U(0, 1)$.
2 If $R_2 \geqslant (1 + R_1^2)^{-1}$ goto 1.
3 generate $R_3 \sim U(0, 1)$.
4 If $R_3 < 0.5$ then
5      $X := R_1$

```
6   Else
7      X := 1/R₁
8   End If
9   Exit.
```

To generate from a full (unfolded) Cauchy distribution, we utilize symmetry, replacing steps 4–9 in Algorithm 3.8 by

```
4    If R₃ < 0.25 then
5       X := R₁
6    Else if R₃ < 0.5 then
7       X := −R₁
8    Else if R₃ < 0.75 then
9       X := 1/R₁
10   Else
11      X := −1/R₁
12   End If.
13   Exit.
```

3.10.2 Standardized Wald distribution

Consider generation from a standardized Wald distribution,

$$f_X(x) = \sqrt{\left(\frac{\beta}{2\pi}\right)} \, x^{-\frac{3}{2}} e^{-\frac{1}{2}\beta(x-1)^2/x} \qquad (x \geq 0)$$

If

$$W = \min(X, X^{-1})$$

eqn (3.37) yields

$$f_W(w) = \sqrt{\left(\frac{\beta}{2\pi}\right)} (w^{-\frac{3}{2}} + w^{-\frac{1}{2}}) e^{-\frac{1}{2}\beta(w-1)^2/w} \qquad (0 \leq w \leq 1).$$

This again has bounded domain, so envelope rejection with a uniform target distribution could be employed. Unfortunately $M = \max[f_W(w)]$ would have to be determined numerically. To avoid this we might consider a further transformation,

$$Y = \beta(W-1)^2/W = \beta(W + W^{-1} - 2).$$

The two transformations are equivalent to

$$Y = g(X) = \beta(X + X^{-1} - 2). \qquad (3.38)$$

Since Y is a decreasing function of W,

$$f_Y(y) = f_W[w(y)] \left|\frac{dw}{dy}\right|$$

$$= \frac{e^{-\frac{1}{2}y} y^{-\frac{1}{2}}}{(\sqrt{2})\Gamma(0.5)} \qquad (y \geq 0).$$

Thus Y is a chi-squared variate, with one degree of freedom. This result is essentially identical to that given by Shuster (1968) for the inverse Gaussian distribution. Since an efficient way of generating such a chi-squared variate is to take the square of a normal deviate, a method based on the transformation (3.38) is promising, and has been suggested by Michael et al. (1976). Given that $Y = y$, the roots of (3.38) are

$$x_1 = 1 + \frac{y}{2\beta} - \frac{(y^2 + 4\beta y)^{\frac{1}{2}}}{2\beta}$$

and

$$x_2 = 1/x_1.$$

From eqn (3.34)

$$p_1(y) = \left\{ 1 + \frac{f_X(x_2)}{f_X(x_1)} \left| \frac{g'_X(x_1)}{g'_X(x_2)} \right| \right\}^{-1}$$

$$= \{1 + x_1(y)\}^{-1}.$$

This gives rise to Algorithm 3.9. Such a method is clearly easy to implement and has zero set-up time. The marginal generation time will depend upon the method of generating Normal deviates. A reasonable compromise between ease of implementation and efficiency is the Box–Müller method, described in Section 4.2.

Algorithm 3.9

Input $[\beta]$
Output $[X]$

1 generate $Z \sim N(0, 1)$, $R \sim U(0, 1)$. $Y := Z^2$.
2 $X := 1 + \{Y - (Y^2 + 4\beta Y)^{\frac{1}{2}}\}/(2\beta)$.
3 If $R > 1/(1 + X)$ then
4      $X := 1/X$
5 End If.
6 Exit.

3.11 Sampling from discrete empirical distributions

Some of the methods already dealt with are applicable to discrete distributions, for example Poisson sampling via a simulation of the Poisson process, and geometric sampling through analytical inversion of the c.d.f. In many simulation studies, it appears that the underlying distribution is empirical, reflecting the raw data, with no attempt being made to fit a standard distribution to the data.[†] In these cases, methods are required for generating variates from tables. Clearly the methods to be developed are also applicable to standard distributions, which can always (if required) be expressed in tabular form. In discussing the available methods we will define

$$p_x = f_X(x)$$

$$q_x = \sum_{j \leqslant x} p_j \qquad (x = 1, 2, \ldots, n), \qquad (3.39)$$

$$q_0 = 0$$

where, without loss of generality, the random variable is defined on the integers $\{1, 2, \ldots, n\}$, and if necessary (i.e., if the probability mass function has infinite domain), the distribution is truncated at some suitable point n.

3.11.1 Sequential search

This is simply the inversion method applied to a c.d.f., represented in tabular form, and results in Algorithm 3.10. The speed of the algorithm is related to the mean number of comparisons (step 2) required, prior to delivery of X. This is given by

$$E = E(X) = \sum_{i=1}^{n} ip_i = n - \sum_{k=1}^{n-1} q_k. \qquad (3.40)$$

Algorithm 3.10

Input $[n, p_1, \ldots, p_n]$ Saved $[q_1, \ldots, q_n]$.
Output $[X]$

S1 $q_1 := p_1$.
S2   For $k := 2, n$
S3 $q_k := q_{k-1} + p_k$
S4   Next k.
1   Generate $R \sim U(0, 1)$, $X := 1$.
2   While $q_X < R$ do
3        $X := X + 1$
4   End While.
5   Exit.

[†] It could be argued that it is preferable to fit a standard distribution to any empirical distribution. From a modelling viewpoint, subsequent experiments involving parametric analysis may be performed simply by altering the parameter values of the standard distribution.

The method can be applied to standard distributions providing a table of cumulative probabilities $\{q_x\}$ is precalculated. In applications involving standard distributions where the parameter values change between calls of the generator, or when it is inconvenient to store a large table of constants, an 'unstored' version based on repeated partial calculation of the c.d.f. is possible. A recursion $p_{j+1} = g_j(p_j)$ can be used to calculate successive cumulative probabilities, as required. Algorithm 3.11 gives the unstored version of 3.10. Note that there is now no need to truncate the distribution at n. A further refinement, known as 'chop-down search' (Kemp, 1981) eliminates the need for the variable s. Instead of incrementing the cumulative probability (s), the random number is decremented. This saves one assignment. Algorithm 3.12 gives this version.

Algorithm 3.11

Input $[p_1, g_{\cdot}(\cdot)]$
Output $[X]$

1   generate $R \sim U(0, 1)$, $X := 1$, $r := p_1$, $s := r$.
2   While $s < R$ do
3      $r := g_X(r)$
4      $X := X + 1$
5      $s := s + r$
6   End While.
7   Exit.

Algorithm 3.12

Input $[p_1, g_{\cdot}(\cdot)]$
Output $[X]$

1   generate $R \sim U(0, 1)$, $X := 1$, $r := p_1$.
2   While $r < R$ do
3      $R := R - r$
4      $r := g_X(r)$
5      $X := X + 1$
6   End While.
7   Exit.

For standard discrete distributions having small means, these unstored versions can give rise to easily implemented algorithms which are efficient, have minimal storage requirements and no set-up time. For distributions having larger means, the number of evaluations of $\{g_X(r)\}$ makes them less attractive. In these cases the search may be initiated at the mode m, finding the smallest X such that $R < q_X$ if $R \geqslant q_m$, or the largest X such that $R \geqslant q_{X-1}$ if $R < q_m$. Algorithm 3.13 shows this procedure, where $p_{j-1} = h_j(p_j)$. Unfortunately some set-up time is now incurred as p_m and q_m have to be calculated and stored. However, for distributions which are not too asymmetric the mean number of comparisons will vary approximately as the standard deviation rather than the mean, which will usually reduce marginal generation time.

Algorithm 3.13

Input $[m, p_m, q_m, g(\cdot), h(\cdot)]$
Output $[X]$

```
1     generate R ~ U(0, 1), X := m, p := pₘ, q := qₘ.
2     If R ≥ q then
3        While R ≥ q do
4           p := g_X(p)
5           X := X + 1
6           q := q + p
7        End While.
8     Else
         q := q - p
9        While R < q
10          p := h_X(p)
11          X := X - 1
12          q := q - p
13       End While.
14    End If.
15    Exit.
```

3.11.2 Ordered sequential search

Expression (3.40) indicates that E may be reduced by a preliminary sort of the integers $\{1, 2, \ldots, n\}$, such that the sorted table of probabilities is non-increasing. This gives rise to Algorithm 3.14. Comparing Algorithms 3.10 and 3.14, $q'_j \geqslant q_j$, with equality for all j, only if the original distribution was non-increasing for all j. Thus, excluding time for the sort, (3.40) indicates that Algorithm 3.14 is no slower, and in general faster than Algorithm 3.10. A disadvantage of this method is that the sort-time will vary as $O(n \ln n)$, making the set-up time potentially large. Kemp (1982) uses a partial sort, searching alternately either side of the mode of the distribution. In effect this is a variant of Algorithm 3.13, with the added advantage that q_m does not have to be calculated.

Algorithm 3.14

Input $[n, p_1, \ldots, p_n]$ Saved $[r_1, \ldots, r_n, q'_1, \ldots, q'_n]$.
Output $[X]$

S1   Find a permutation $\{r_1, \ldots, r_n\}$ of $\{1, \ldots, n\}$
     such that $p_{r_1} \geqslant \ldots \geqslant p_{r_n}$.
S2 $q'_1 := p_{r_1}$.
S3   For $k := 2, n$
S4 $q'_k := q'_{k-1} + p_{r_k}$
S5   Next k.
1   Generate $R \sim U(0, 1)$, $X := 1$.
2   While $q'_X < R$ do
         $X := X + 1$
3   End While.
4    $X := r_X$.
5   Exit.

3.11.3   Indexed search

This is a refinement of sequential search, due to Chen and Asau (1974). The probability distribution is first divided into k equal parts ($k = 1, 2, \ldots$). A random number is used to determine in which of the k intervals the generated value lies, then the selected interval is searched for the correct value. Note that this is a natural extension of sequential search, starting at the mode.

Given a distribution with probabilities $(p_1, p_2, \ldots, p_n)$ and cumulative probabilities $(q_1, q_2, \ldots, q_n)$, intervals (s_{j-1}, s_j) $(j = 1, 2, \ldots, k)$ are constructed where s_j is the smallest integer satisfying

$$q_{s_j} \geqslant \frac{j}{k}$$

and

$$s_0 = 1.$$

Algorithm 3.15 then generates variates from the distribution.

Algorithm 3.15

Input $[k, s_0, \ldots, s_k, q_1, \ldots, q_n]$
Output $[X]$

1   Generate $R \sim U(0, 1)$.
2   $j := \langle kR + 1 \rangle$.
3   Find smallest $i \in (s_{j-1}, s_j)$ such that
        $q_i \geqslant R$.
4   $X := i$.
5   Exit.

To illustrate the method, consider the 10-point distribution having cumulative probabilities: $q_1 = 0.03$, $q_2 = 0.17$, $q_3 = 0.48$, $q_4 = 0.65$, $q_5 = 0.85$, $q_6 = 0.95$, $q_7 = 0.97$, $q_8 = 0.98$, $q_9 = 0.99$, $q_{10} = 1.00$. If we employ $k = 5$ intervals, the boundaries are $s_0 = 1$, $s_1 = 3$, $s_2 = 3$, $s_3 = 4$, $s_4 = 5$, $s_5 = 10$. If, for example, a random number 0.7234 is generated then $\langle kR+1 \rangle = \langle 4.6170 \rangle = 4$, directing the search to the interval $(s_3, s_4) = (4, 5)$. The smallest i, such that $q_i \geq 0.7234$ necessarily lies in this interval and is 5. Thus $X = 5$ is delivered.

There is a trade-off between the number of intervals and the mean number of comparisons within an interval. As k becomes large, the latter approaches a limiting value of 1.

3.11.4   Marsaglia's method

The previous methods can be speeded up, at the cost of increased storage, by a scheme due to Marsaglia (1963). Suppose the probabilities $\{p_x\}$ are expressed to m decimal places, and are all non-zero. Define,

$$A(j) = \begin{cases} 1 & (0 < j \leq 10^m q_1) \\ 2 & (10^m q_1 < j \leq 10^m q_2) \\ \vdots & \\ n & (10^m q_{n-1} < j \leq 10^m q_n). \end{cases}$$

Then Algorithm 3.16 generates variates from the probability distribution.

Algorithm 3.16

Input $[m, A(\cdot), \ldots, A(10^m)]$
Output $[X]$

1   Generate $R \sim U(0, 1)$.
2   $X := A(\langle 10^m R \rangle + 1)$.
3   Exit.

The validity follows immediately, as

$$P(X = i) = P(10^m q_{i-1} < \langle 10^m R \rangle + 1 \leq 10^m q_i)$$

$$= \frac{10^m q_i - 10^m q_{i-1}}{10^m}, \tag{3.41}$$

since $\langle 10^m R \rangle + 1$ is uniformly distributed on the integers $\{1, 2, \ldots, 10^m\}$. Thus

$$P(X = i) = q_i - q_{i-1} = p_i,$$

as required.

The algorithm is fast, but rapidly becomes impracticable since 10^m storage locations are required.

3.11.5 Method of Norman and Cannon

This was first outlined by Marsaglia (1963), and later implemented by Norman and Cannon (1972). For ease of exposition the method is explained first with the aid of a small example. Subsequently an algorithm will be given and verified, as no formal proof is given by Norman and Cannon.

Consider a distribution with probabilities $p_1 = 0.356$, $p_2 = 0.279$, and $p_3 = 0.365$. Each probability may be split into its component digits as follows:

$p_1 = 0.3$	5	6
$p_2 = 0.2$	7	9
$p_3 = 0.3$	6	5
8	18	20

Each probability is expressed to three decimal places, and the entries in the fourth row give the column sums for each decimal place. An array $A(j)$ containing 46 elements is set up. Of the first eight elements, the first three are filled with 1s, the next two with 2s, and the last three with 3s. Of the next 18 elements, the first five are filled with 1s, the next seven with 2s and the last six with 3s. Of the final 20 elements, the first six are filled with 1s, the second nine with 2s and the last five with 3s.

Given a random number R, if $\langle 10R \rangle < 8$, then the random variate is delivered as $A(\langle 10R \rangle + 1)$. The effect of this is that, conditional upon $0 \leqslant R < 0.8$, the numbers 1, 2, and 3 are returned with probabilities $\frac{3}{8}, \frac{2}{8}$, and $\frac{3}{8}$, respectively. If the first test fails, then a check is made to establish whether $8 \times 10 \leqslant \langle 100R \rangle < 8 \times 10 + 18$. If so, then the random variate is delivered as $A(8 + \langle 100R \rangle - 80 + 1)$. Thus conditional upon $0.8 \leqslant R < 0.98$, the probabilities of obtaining 1, 2, and 3 are $\frac{5}{18}, \frac{7}{18}$, and $\frac{6}{18}$, respectively. If this test fails, it must be the case that $8 \times 100 + 18 \times 10 \leqslant \langle 1000R \rangle < 8 \times 100 + 18 \times 10 + 20 = 1000$, in which case the random variable is delivered as $A(26 + \langle 1000R \rangle - 980 + 1)$. Thus, conditional upon $0.98 \leqslant R < 1$, the probabilities of obtaining 1, 2, and 3 are $\frac{6}{20}, \frac{9}{20}$, and $\frac{5}{20}$, respectively. Using this procedure we can, for example, deduce the probability that the random variate $X = 1$:

$$P(X = 1) = \tfrac{3}{8} P(\langle 10R \rangle < 8) + \tfrac{5}{18} P(80 \leqslant \langle 100R \rangle < 98)$$
$$+ \tfrac{6}{20} P(980 \leqslant \langle 1000R \rangle < 1000)$$
$$= \tfrac{3}{8}(0.8) + \tfrac{5}{18}(0.18) + \tfrac{6}{20}(0.02)$$
$$= 3 \times 10^{-1} + 5 \times 10^{-2} + 6 \times 10^{-3} = 0.356,$$

as required.

To formalize the procedure, suppose p_i is expressed to m decimal places:

$$p_i = 0 \cdot \delta_{i1} \delta_{i2} \ldots \delta_{im} \qquad (i = 1, 2, \ldots, n).$$

Define,

$$\left.\begin{array}{c} \delta_{0j} = 0 \\[6pt] n_j = \sum_{i=1}^{n} \delta_{ij} \end{array}\right\} \quad (j = 1, 2, \ldots, m)$$

$$n_0 = 0,$$

$$f_j = n_j + 10f_{j-1} \quad (j = 1, 2, \ldots, m)$$

and   (3.42)

$$f_0 = 0.$$

Define the array $A(\cdot)$ by,
$$A(s) = i,$$

if there exists a j, such that

$$\sum_{k=0}^{i-1} \delta_{kj} < s - \sum_{l=0}^{j-1} n_l \leqslant \sum_{k=0}^{i} \delta_{kj}, \qquad (3.43)$$

for $s = 1, 2, \ldots, \sum_{j=1}^{m} n_j$, and $i = 1, 2, \ldots, n$.

Then Algorithm 3.17 generates variates from the n-point distribution having probabilities $p_1, p_2, \ldots, p_n$.

Algorithm 3.17

Input $\left[m, f_0, \ldots, f_m, n_0, \ldots, n_m, A(1), \ldots, A\left(\sum_{j=1}^{m} n_j\right) \right]$

Output $[X]$

1 generate $R \sim U(0, 1)$, $j := 0$.
2 Repeat
3    $j := j + 1$
4 Until $f_j - n_j \leqslant \langle 10^j R \rangle < f_j$.
5 $X := A\left(\sum_{l=0}^{j-1} n_l + \langle 10^j R \rangle - f_j + n_j + 1 \right)$.
6 Exit.

To verify the procedure, we must first show that the events

$$f_j - n_j \leqslant \langle 10^j R \rangle < f_j \qquad (j = 1, 2, \ldots, m) \qquad (3.44)$$

are mutually exclusive. Repeated application of (3.42) gives

$$f_j = n_j + 10n_{j-1} + \ldots + 10^{j-1}n_1.$$

Thus (3.44) becomes

$$10^{-j}(10n_{j-1} + \ldots + 10^{j-1}n_1) \leqslant R < 10^{-j}(n_j + 10n_{j-1} + \ldots + 10^{j-1}n_1)$$

or

$$10^{-1}n_1+\ldots+10^{-(j-1)}n_{j-1}\leqslant R<10^{-1}n_1+\ldots+10^{-(j-1)}n_{j-1}+10^{-j}n_j. \tag{3.45}$$

As j runs over the integers 1 to m, we see that (3.45), and hence (3.44), generates non-overlapping intervals spanning the interval $[0, 1)$.

Because of this mutual exclusivity, we have from step 5 of Algorithm 3.17 that

$$P(X = i) = \sum_{j=1}^{m} P\left\{A\left(\sum_{l=0}^{j-1} n_l + \langle 10^j R\rangle - f_j + n_j + 1\right) = i\right\},$$

which in conjunction with (3.43) gives

$$P(X=i) = \sum_{j=1}^{m} P\left\{\sum_{k=0}^{i-1} \delta_{kj} + f_j - n_j < \langle 10^j R\rangle +1 \leqslant \sum_{k=0}^{i} \delta_{kj} + f_j - n_j\right\}. \tag{3.46}$$

Since $\langle 10^j R\rangle +1$ is uniformly distributed on the integers $\{1, 2, \ldots, 10^j\}$ (3.46) becomes

$$P(X = i) = \sum_{j=1}^{m} 10^{-j}\left\{\sum_{k=0}^{i} \delta_{kj} - \sum_{k=0}^{i-1} \delta_{kj}\right\}$$

$$= \sum_{j=1}^{m} 10^{-j}\delta_{ij}$$

$$= 0\cdot\delta_{i1}\delta_{i2}\ldots\delta_{im},$$

as required.

The main features of Algorithm 3.17 are that it is fast and has low storage requirements (compared to Marsaglia's method). However, it requires some time to set up the array $A(\cdot)$. The reduction in storage requirements over Marsaglia's method can be significant. By way of an illustrative example, consider the binomial distribution with parameters $n = 10$ and $p = 0.3$. Expressed to four decimal places, the probabilities are $p_0 = 0.0282$, $p_1 = 0.1211$, $p_2 = 0.2335$, $p_3 = 0.2668$, $p_4 = 0.2001$, $p_5 = 0.1030$, $p_6 = 0.0367$, $p_7 = 0.0090$, $p_8 = 0.0015$, $p_9 = 0.0001$ and $p_{10} = 0.0000$. Using Marsaglia's method an array of size 10^4 is required, whereas the Normal and Cannon implementation requires $\sum_{j=1}^{4} n_j = 8 + 16 + 37 + 30 = 91$ storage locations.

Finally, we may recall a remark made in Section 1.5.1, namely that, with a few exceptions, exact algorithms would be considered. The Marsaglia and Norman–Cannon methods are both examples of approximate methods, since the probabilities have been expressed to m decimal places. Of course, this occurs in any implementation of a random variate generator, due to the finite word length of the computer. However, in the case of Marsaglia's method the effect may not be insignificant, since m will have to be made reasonably small in order to contain the array size.

4

Methods of generation from specific continuous distributions

In Chapter 3 we discussed a number of generation methods. For a given distribution some of these may be inappropriate, others may work poorly, and some may be particularly suitable. Given the number of general methods available and the potential flexibility offered by each (for example the choice of target distribution within envelope rejection) it is not surprising to find that common distributions such as the normal, gamma, and beta now have a very large number of alternative generation procedures. For example, there are at least 20 documented methods for generating gamma variates. An attempt will be made to describe or mention most methods for the sake of completeness. However, the reader is probably also interested in identifying the 'best' gamma variate generator. Ideally the selection of a 'best' algorithm would involve some reference to the five attributes described in Section 1.5. In practice this is impossible without a knowledge of the user's utility function for these five attributes! It would also require experimental evidence for the behaviour of all methods on a number of different machines. Clearly such an approach is verging on the impossible. Even if it were feasible it is debatable whether the resources used in such a study could be justified, with respect to the conclusions reached. Fortunately, when we are dealing with families of distributions, such as the beta or gamma, we may narrow down the choice by insisting that a really useful generator must be efficient over a wide range of parameter values. This is where a theoretical analysis of specific algorithms can be useful. For example, when we considered in Section 3.6 the standard ratio of uniforms method applied to a gamma distribution, the acceptance probability was found to be $\sqrt{\{\pi/(2\alpha)\}}$, for large α. Thus this algorithm has an unbounded marginal generation time with respect to the shape parameter α. Given the choice of gamma algorithms available, we would be unlikely to implement such a method, and would be justified in narrowing down the search to a method with bounded generation time. Following this rationale the description which follows tends to be more heavily weighted in favour of those algorithms which can be shown theoretically to have a robust performance with respect to the parameters of the distribution.

4.1 Negative exponential

Generation from the negative exponential distribution is important not only
in its own right, but also because other generators frequently require a source
of exponential variates. Y has a negative exponential distribution if its p.d.f. is

$$f_Y(y) = \lambda e^{-\lambda y} \qquad (y \geqslant 0),$$

where $\lambda > 0$. The distribution may be standardized by setting $X = \lambda Y$, giving

$$f_X(x) = e^{-x} \qquad (x \geqslant 0).$$

Perhaps the most familiar method for generating X-values is to invert the
c.d.f. This gives $X = -\ln R$.

An alternative technique uses Forsythe's method. Using the notation of
Section 3.7,

$$\left. \begin{array}{l} \psi_j(x) = e^{-(x-[j-1])}/(1-e^{-1}) \\ h_j(x) = x-[j-1] \end{array} \right\} \qquad (j-1 \leqslant x < j),$$

$$p_j = e^{-(j-1)}(1-e^{-1}), \tag{4.1}$$

for $j = 1, 2, \ldots$. In this instance there is no necessity to calculate and store
$\{p_j\}$. This follows since the probability that a random number comparison
sequence terminates in acceptance of the prospective variate is $1 - e^{-1}$. Thus
the probability that the jth prospective variate is accepted is p_j. Further,
sampling from $\psi_j(x)$ is equivalent to sampling from $\psi_1(x)$ and adding $(j-1)$.
The resulting method is to generate sequences of random numbers $\{R^{(k)}, R_1^{(k)},
R_2^{(k)}, \ldots\}$, $(k = 1, 2, \ldots)$, stopping at $k = j$, when $N^{(k)}$ is odd-valued, where

$$N^{(k)} = 1 \quad \text{if } R^{(k)} < R_1^{(k)},$$

and $N^{(k)} = n \quad \text{if } R^{(k)} \geqslant R_1^{(k)} \geqslant \ldots \geqslant R_{n-1}^{(k)} < R_n^{(k)}, (n \geqslant 2).$

$X = j - 1 + R^{(j)}$ is then delivered as the sample variate. The mean number of
random numbers required is $e^2/(e-1) = 4.30$. Apart from this, only comparisons and additions are involved. Algorithm E1 shows how the method
may be implemented.

Some timing results may be of interest here. On a DEC-20 using the
FORTRAN random number function RAN (13 μs per uniform variate), the
logarithmic method took 64 μs, while Forsythe's method took 73 μs. Both
implementations were FORTRAN based. Ripley (1983a) using FORTRAN
on a Cyber 174 found generation time to be 80 and 50 μs, respectively, using
the fast RANF uniform generator (2.8 μs). However, the position was re-
versed (100 and 180 μs) when the NAG random number function G05CAF
(30 μs) was used on the same machine. Ahrens and Dieter (1972) found
FORTRAN implementations of the logarithmic method (LG) and Forsythe's
method (NE) to take 503 and 756 μs, respectively, on an IBM 360/50, when

Algorithm E1

Output [X]

1     $j:=0$.
2   generate $R \sim U(0, 1)$, $E:=R$.
3   generate $R_1 \sim U(0, 1)$.
4   If $R_1 > R$ then
5       $X:=j+E$
6   Exit
7   End If.
8   generate $R \sim U(0, 1)$.
9   If $R \leqslant R_1$ goto 3.
10    $j:=j+1$, goto 2.

the uniform generator took 105 μs. An assembler implementation of NE (with a machine-coded uniform generator) took one-third of the FORTRAN time, but no corresponding times are available for LG. These variations indicate that it is probably worthwhile to determine which of the two methods is faster in the user's particular environment. In other respects the methods are very similar, requiring no set-up calculations or storage of constants. Both can be rapidly implemented.

A method due to Marsaglia (1961) samples from

$$\psi_1(x) = e^{-x}/(1 - e^{-1}) \qquad (0 \leqslant x \leqslant 1),$$

by generating a random variable,

$$Y = \min(R_1, R_2, \ldots, R_M), \tag{4.2}$$

where M is a random variable having p.m.f.

$$f_M(m) = [(e-1)m!]^{-1} \qquad (m = 1, 2, \ldots).$$

The sample variate is then delivered as $X = Y + j - 1$, where j is sampled from (4.1). The method involves look-up tables for M and j, and is clearly not as elegant as Forsythe's. Ahrens and Dieter's FORTRAN implementation (MA) was marginally slower than NE on an IBM 360/50, while the assembler implementation was slightly faster.

Another approach by Ahrens and Dieter (1972) (SA) relies implicitly on generation from the p.d.f.,

$$f_X(x) = (\ln 2)e^{-x \ln 2} \qquad (x \geqslant 0),$$

which may be represented as the probability mixture

$$f_X(x) = \sum_{j=1}^{\infty} p_j \psi_j(x),$$

where $$p_j = 2^{-j} \tag{4.3}$$

and $$\psi_j(x) = (\ln 2)2^{-(x-j)} \qquad (j-1 \leqslant x < j),$$

for $j = 1, 2, \ldots$. For fixed j, sampling from $\psi_j(x)$ is performed by sampling from $\psi_1(x)$ and adding $j-1$ to the result. Sampling from (4.3) is performed by examining a sequence of independent bits taking value 0 or 1 with equal probability. Ahrens and Dieter found that a FORTRAN implementation was approximately 10% slower than LG, while an assembler implementation was faster than MA and NE.

A method which has the reputation of being very fast on most machines is due to Maclaren *et al.* (1964). They use the composition method on

$$f_X(x) = \sum_{i=1}^{3} a_i g_i(x),$$

where

$$\left.\begin{array}{ll} g_1(x) = a_1^{-1} e^{-(k+1)c} \\ g_2(x) = a_2^{-1} (e^{-x} - e^{-(k+1)c}) \end{array}\right\} \qquad \begin{array}{l} (kc \leqslant x < (k+1)c) \\ (k = 0, 1, \ldots, 4/c - 1) \end{array}$$

$$g_3(x) = a_3^{-1} e^{-x} \qquad\qquad (4 \leqslant x),$$

$$a_1 = c \sum_{k=0}^{(4/c)-1} e^{-(k+1)c} = c(1 - e^{-4})/(e^c - 1),$$

$$a_3 = e^{-4},$$

$$a_2 = 1 - a_1 - a_3,$$

and c is chosen such that $(4/c)$ is integer. c was chosen to be $1/16$ for a binary machine and 0.1 for a decimal machine. $g_1(x)$ is a piecewise uniform distribution on $[0, 4)$ and sampling needs one random number plus an appropriate table of cumulative probabilities. Note that most of the probability mixture is concentrated into $g_1(x)$, where sampling is easiest. In the region $[0, 4)$ $g_2(x)$ represents the (small) difference between $f_X(x)$ and $g_1(x)$. It has the appearance of a series of wedges. Generation is via a discrete random variate (to determine which wedge) and a continuous variate, the latter constructed from the minimum of Z random numbers, where Z has a known distribution (this is analogous to the generation of Y in eqn (4.2)). Sampling from $g_3(x)$ is accomplished by adding 4 to an exponential variate. The latter is obtained by re-entering the whole procedure (and this process continues until eventually sampling is effected from $g_1(x)$ or $g_2(x)$). The method is efficient in the sense that no time-consuming function evaluations are required and few random numbers are needed. The mean requirement of random numbers is $(1 + ha_2)$ $/(a_1 + a_2)$, where h is the expected number needed for sampling from $g_2(x)$. For $c = 1/16$ this gives the overall mean as 1.11 random numbers (Fishman, 1978, p. 410).

In assessing the suitability of methods for the exponential, the available computational evidence suggests that for FORTRAN implementations the logarithmic method is often not substantially slower than the others, and sometimes faster. For programming in machine code, comparative results involving the logarithmic method are not readily available. One advantage of the method is that by setting $X_1 = -\ln R$ and $X_2 = -\ln(1-R)$, high-quality antithetic variates may be obtained. This gives the bivariate exponential distribution considered in Section 1.1. The correlation between X_1 and X_2 is -0.645 (Fishman, 1973, p. 320). For a machine with an extremely fast random number generator, Forsythe's method may be faster. The storage of a large number of constants, and the effort required to code the algorithm, appear to be the main disadvantages of Maclaren et al.'s ingenious method. The logarithmic method appears as a FORTRAN 77 function EXPON in Appendix 1.5.

4.2 Normal distribution

A well-known, if inadvisable, method of generating standard Normal deviates is to take 12 random numbers, $\{R_i : i = 1, 2, \ldots, 12\}$ and set $X = \sum_{i=1}^{12} R_i - 6$. By the Central Limit Theorem, the distribution of X is approximately standard normal. This method should not be used. It is an approximation only, and the fit is poor in the tails of the distribution. Since simulations are frequently undertaken to examine the response of a system to rare events, this may be serious. Further, the requirement for 12 uniform deviates is likely to make such a method slow, even when random number generation is fast.

Sections 4.2.1–4.2.4 describe exact methods for Normal generation, which are easy to implement (i.e. require few lines of code), require no storage of constants, and little if any set-up time. Sections 4.2.5–4.2.7 describe methods complementary to the first set. In these cases, the methods tend to require more programming effort, often require the storage of a large number of constants, and occasionally involve set-up calculations.

4.2.1 Box–Müller method

Given two random numbers R_1 and R_2, a pair of standard Normal deviates can be obtained via

$$X_1 = (-2\ln R_1)^{\frac{1}{2}} \cos(2\pi R_2)$$
$$X_2 = (-2\ln R_1)^{\frac{1}{2}} \sin(2\pi R_2).$$

(4.4)

The proof, due to Box and Müller (1958) is as follows. Let X_1 and X_2 be a pair of independent standard normal deviates. Consider a transformation $(X_1, X_2) \to (V, \Theta)$ where

$$X_1 = V \cos \Theta; \quad X_2 = V \sin \Theta. \tag{4.5}$$

The Jacobian of this transformation is V, and hence the joint p.d.f. of V and Θ is

$$f_{V, \Theta}(v, \theta) = \frac{e^{-\frac{1}{2}v^2} v}{2\pi} \quad (v \geqslant 0, 2\pi \geqslant \theta \geqslant 0).$$

Thus Θ and V are independently distributed with $\Theta \sim U(0, 2\pi)$ and $\frac{1}{2}V^2 \sim E$, where E denotes an exponential variate, mean 1. Using the inversion method, these two random variates may be generated by

$$\Theta = 2\pi R_2$$

and

$$\tfrac{1}{2}V^2 = -\ln R_1.$$

The result (4.4) follows on applying the transformation (4.5).

Although the method is clearly exact from a mathematical viewpoint, the distribution of the sample values may be seriously different from that expected, when certain multiplier values are used in a linear congruential random number generator (multiplicative or mixed). In an empirical investigation Neave (1973) generated 10^6 standard normal deviates using the sine part only of result (4.4), together with a multiplicative generator with multiplier $a = 131$ and modulus $m = 2^{35}$. Disturbingly, no deviates were obtained outside the range $(-3.3$ to $3.6)$. The explanation for this is that a tail deviate can be obtained only if R_1 is small, but R_2 is not so. However, for a multiplier as low as 131, a small value of R_1 always leads to a small value of R_2. A similar effect occurs with mixed generators. In both cases the effect becomes less pronounced as a becomes larger. In eqn (4.4) it is understood that the random number R_1 is generated before R_2. Chay et al. (1975) suggested exchanging R_1 and R_2. The effect of this is identical to that obtained with no exchange, but using a multiplier a^*, where $aa^* = 1 \bmod m$ (a^* is the multiplier that would generate the original stream in reverse order). For Neave's example a^* takes the value 16 261 860 907. Since the multiplier is now very much larger, the Neave effect becomes less pronounced with deviates attainable in the range $(-6.8992$ to $7.0576)$. In general, however, the effect is worrying, since the experimenter will often not know the value of the multiplier. If the method is used it is safer to shuffle the output from the random number generator first (see Section 2.3.3).

Using the Box–Müller method, normal deviates have to be generated in pairs. One is used and the other stored for subsequent use. It is an easily implemented method requiring one square root, one logarithmic, and two trigonometric evaluations per pair of deviates. On some computers these evaluations may slow the procedure down.

A modification due to Marsaglia and Bray (1964) avoids the trigonometric functions and will often, but not always speed up the procedure. We shall refer to this as the *polar* Box–Müller method. Given two independent random variables U_1, $U_2 \sim U(-1, 1)$, if $U_1^2 + U_2^2 \leqslant 1$, then $\tan^{-1}(U_2/U_1)$ and $U_1^2 + U_2^2$ are independently distributed $\sim U(0, 2\pi)$ and $\sim U(0, 1)$, respectively. Setting $R_1 = U_1^2 + U_2^2$ and $2\pi R_2 = \tan^{-1}(U_2/U_1)$, eqns (4.4) become

$$X_1 = \{-2\ln(U_1^2 + U_2^2)\}^{\frac{1}{2}} U_1 (U_1^2 + U_2^2)^{-\frac{1}{2}}$$

$$X_2 = \{-2\ln(U_1^2 + U_2^2)\}^{\frac{1}{2}} U_2 (U_1^2 + U_2^2)^{-\frac{1}{2}}.$$

A mean of $4/\pi$ uniform deviates is required per variate. This form does not appear to suffer from the problems reported by Neave, and is therefore preferred to the basic Box–Müller method. Algorithm N1 gives the polar form. I, B, and U_2 are variables whose values are saved between calls of the generator. I is an indicator (initially set to zero) taking value 1 or 0 according to whether one deviate from the original pair is currently stored or not.

Algorithm N1

Output $[X]$ Saved $[I, B, U_2]$.

S1   Data : $I/0/$
1   If $I = 1$ then
2         $X := BU_2$
3   Else
4   generate $Y_1, Y_2 \sim U(0, 1)$
5         $U_1 := 2Y_1 - 1$
6         $U_2 := 2Y_2 - 1$
7         $S := U_1^2 + U_2^2$
8   If $S \geqslant 1$ goto 4
9         $B := \{-2\ln S/S\}^{\frac{1}{2}}$
10       $X := BU_1$
11   End If.
12   $I := (I + 1) \bmod 2.$
13   Exit.

4.2.2   Butcher's method

Another early generation method is due to Butcher (1961). Generation is from a folded normal distribution,

$$f_X(x) = \left(\frac{2}{\pi}\right)^{\frac{1}{2}} e^{-\frac{1}{2}x^2} \qquad (x \geqslant 0). \tag{4.6}$$

Envelope rejection is used with a target distribution

$$g_Y(x) = \begin{cases} \frac{2}{3} & (0 \leqslant x \leqslant 1) \\ \frac{2}{3}e^{-2(x-1)} & (x > 1). \end{cases} \qquad (4.7)$$

Figure 4.1 illustrates how the normal density is enveloped.

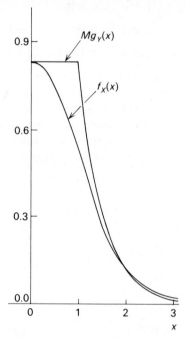

Fig. 4.1. Butcher's envelope rejection method for the normal distribution

Using inversion, Y-values are generated by

$$Y = \begin{cases} \frac{3}{2}R_1 & (R_1 \leqslant \frac{2}{3}) \\ 1 - \frac{1}{2}\ln(3 - 3R_1) & (R_1 > \frac{2}{3}). \end{cases} \qquad (4.8)$$

From (4.6) and (4.7),

$$\frac{f_X(x)}{g_Y(x)} = \begin{cases} \left(\frac{3}{2}\right)\left(\frac{2}{\pi}\right)^{\frac{1}{2}} e^{-\frac{1}{2}x^2} & (0 \leqslant x \leqslant 1) \\ \left(\frac{3}{2}\right)\left(\frac{2}{\pi}\right)^{\frac{1}{2}} e^{-\frac{1}{2}(x-2)^2} & (x > 1) \end{cases}$$

and

$$M = \max\left[f_X(x)/g_Y(x)\right] = \left(\frac{3}{2}\right)\left(\frac{2}{\pi}\right)^{\frac{1}{2}} = 1.197.$$

Thus the acceptance condition becomes

$$R_2 < \begin{cases} e^{-\frac{1}{2}Y^2} & (0 \leqslant Y \leqslant 1) \\ e^{-\frac{1}{2}(Y-2)^2} & (Y > 1) \end{cases}$$

or

$$E = -\ln R_2 > \begin{cases} \frac{1}{2}Y^2 & (0 \leqslant Y \leqslant 1) \\ \frac{1}{2}(Y-2)^2 & (Y > 1), \end{cases}$$

where E is a standard exponential variate. Algorithm N2 shows how the method may be implemented. The number of logarithmic evaluations is reduced considerably by using the 'squeeze' technique (Section 3.4). We use the bounds $R_2^{-1} - 1 \geqslant -\ln R_2 \geqslant 1 - R_2$ to derive acceptance and rejection pretests (steps 9 and 10). The M-value gives the acceptance probability as 0.836, the mean number of uniform deviates required as 3.39, and the mean number of logarithmic evaluations (in a version without pretests) as 1.60. The pretest version given here reduces that number to approximately 0.454. The latter is a sample value based on the generation of 10 000 variates. At the expense of slightly more code the number of uniform deviates can be reduced to 2.40, by redefining $g_Y(\cdot)$ so that $Y \in (-\infty, \infty)$ and $g_Y(\cdot)$ is symmetric about the origin.

Algorithm N2

Output [X]

```
1    generate R₁, R₂ ~ U(0, 1).
2    If R₁ ≤ ⅔ then
3        X := ³⁄₂R₁
4        W := ½X₁²
5    Else
6        X := 1 - ½ln(3 - 3R₁)
7        W := ½(X - 2)²
8    End If.
9    If 1 - R₂ > W goto 12.
10   If 1/R₂ - 1 ≤ W goto 1.
11   If -ln R₂ ≤ W goto 1.
12   generate R₃ ~ U(0, 1).
13   If R₃ ≥ 0.5 then
14       X := -X
15   End If.
16   Exit.
```

4.2.3 Band rejection method

Payne (1977) has described an elegant method using the band rejection technique. The folded normal distribution is expressed as a probability mixture,

$$f_W(w) = 0.9545\, f_X(x) + 0.0455\, f_Z(x),$$

where

$$f_X(x) = \left(\frac{2}{\pi}\right)^{\frac{1}{2}} e^{-\frac{1}{2}x^2} \Big/ 0.9545 \qquad (0 \leqslant x \leqslant 2)$$

$$f_Z(x) = \left(\frac{2}{\pi}\right)^{\frac{1}{2}} e^{-\frac{1}{2}x^2} \Big/ 0.0455 \qquad (x > 2).$$

Sampling from $f_Z(\cdot)$ is via a tail generation procedure due to Marsaglia (1964). This is explained in more detail in Section 7.3. For the moment we note that if R_1, $R_2 \sim U(0, 1)$ and $R_2 < 2/\sqrt{(4 - 2\ln R_1)}$, then $Z = \sqrt{(4 - 2\ln R_1)}$. The probability of such an acceptance is 0.8425.

The main part of the procedure is concerned with generating X-values. Since most of the probability mixture is concentrated here, a high sampling efficiency is required for this part. Using a uniform target distribution,

$$g_Y(x) = 0.5 \qquad (0 \leqslant x \leqslant 2),$$

$$\frac{f_X(x)}{g_Y(x)} = \left(\frac{2}{\pi}\right)^{\frac{1}{2}} \frac{2}{0.9545} e^{-\frac{1}{2}x^2},$$

$$M = \max[\{f_X(x) + f_X(2 - x)\}/g_Y(x)]$$

$$= \left(\frac{2}{\pi e}\right)^{\frac{1}{2}} \frac{4}{0.9545} = 2.025,$$

giving a sampling efficiency of $2M^{-1}$ or 0.9862. This high value is confirmed by Fig. 4.2, which illustrates the acceptance regions. The primary and secondary acceptance conditions (steps 2 and 5 of Algorithm 3.4) are

$$-2\ln 2R_2 > Y^2 - 1$$

and

$$-2\ln\{2(1 - R_2)\} > (2 - Y)^2 - 1,$$

respectively.

A version of the method is shown in Algorithm N3. Step 8 is the primary acceptance condition and is preceded by pretests for primary acceptance and rejection (steps 6 and 7). Step 12 is the secondary rejection condition and is preceded by an acceptance pretest (step 11). Note that a rejection pretest is not justified since the overall probability of rejection is only 0.0455. The length of the algorithm indicates that it is not so rapidly implemented as the polar Box–Müller or Butcher's methods. A mean of 4.04 random numbers is required per variate, compared with 1.27 and 3.40 for the other methods. One

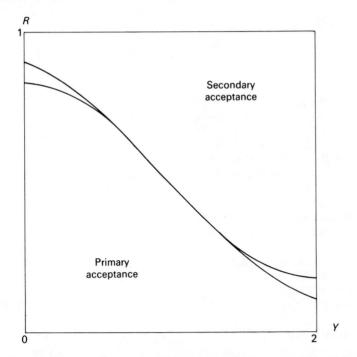

Fig. 4.2. Acceptance regions for band rejection applied to a normal distribution

advantage, however, is that logarithmic evaluations are never needed for generating prospective variates from $f_X(\cdot)$, and are reduced overall by introduction of pretests. A sample of 10 000 variates indicated a mean requirement of only 0.46 logarithmic evaluations per variate.

4.2.4 Ratio of uniforms method

As shown in Section 3.6, the general ratio of uniforms method (with a rectangular enclosing region) is to generate (U, V) uniformly over $D \equiv \{(u, v): 0 \leqslant u \leqslant u_+, -v_- \leqslant v \leqslant v_+\}$, and to deliver V/U if $U \leqslant f_X^{\frac{1}{2}}(V/U)$. In this case the p.d.f.s of X and $Y = X^{-1}$ are

$$f_X(x) = (2\pi)^{-\frac{1}{2}} e^{-\frac{1}{2}x^2},$$

$$f_Y(y) = (2\pi)^{-\frac{1}{2}} y^{-2} e^{-\{1/(2y^2)\}}.$$

$f_X(\cdot)$ is maximized at $m_X = 0$, $f_Y(\cdot)$ at $m_{Y_1, Y_2} = \pm 1/\sqrt{2}$, giving

$$u_+ = f_X^{\frac{1}{2}}(m_X) = (2\pi)^{-\frac{1}{4}},$$

$$v_+ = f_Y^{\frac{1}{2}}(m_{Y_1}) = (2\pi)^{-\frac{1}{4}}(2/e)^{\frac{1}{2}},$$

$$v_- = f_Y^{\frac{1}{2}}(m_{Y_2}) = v_+.$$

Algorithm N3

Output $[X]$

1   generate $R \sim U(0, 1)$.
2   If $R \leqslant 0.9545$ then
3   generate $R_1, R_2 \sim U(0, 1)$
4       $X := 2R_1$
5       $W := \frac{1}{2}(X^2 - 1)$
6   If $1 - \ln 2 - R_2 > W$ goto 18
7   If $1/R_2 - 1 - \ln 2 \leqslant W$ goto 9.
8   If $-\ln 2 - \ln R_2 > W$ goto 18
9       $X := 2 - X$
10      $W := \frac{1}{2}(X^2 - 1)$
11   If $-\ln 2 + R_2 > W$ goto 18
12   If $-\ln 2 - \ln(1 - R_2) \leqslant W$ goto 3.
13   Else
14   generate $R_1, R_2 \sim U(0, 1)$
15      $X := (4 - 2 \ln R_1)^{\frac{1}{2}}$
16   If $R_2 > 2/X$ goto 14
17   End If.
18   generate $R_3 \sim U(0, 1)$.
19   If $R_3 \geqslant 0.5$ then
20      $X := -X$
21   End If.
22   Exit.

Thus the prospective variate V/U is

$$X = \frac{(2R_2 - 1)v_+}{R_1 u_+}$$

and the acceptance condition is

$$R_1 u_+ \leqslant f_X^{\frac{1}{2}}(X)$$

or
$$-\ln R_1 \geqslant \tfrac{1}{4}X^2.$$

To reduce logarithmic evaluations, we may employ the squeeze technique. A simple acceptance pretest is $1 - R_1 \geqslant \frac{1}{4}X^2$, while Knuth (1981, p. 127) suggests $(cR_1)^{-1} - 1 + \ln c < \frac{1}{4}X^2$ for rejecting X. c can·be any value greater than 0. Computational evidence suggests that $c = e^{1.35}$ gives good performance. This gives rise to algorithm N4.

From eqn (3.10) the acceptance probability is $(4u_+ v_+)^{-1}$ or $\sqrt{(\pi e)}/4 = 0.731$. This is not as large as one might hope for, but it should be realized that the frequency of the most time-consuming operation (logarithmic evalu-

Algorithm N4

Output $[X]$

1   generate $R_1, R_2 \sim U(0, 1)$.

2    $X := \left(\dfrac{2}{e}\right)^{\frac{1}{2}} \left(\dfrac{2R_2 - 1}{R_1}\right)$.

3    $W := \frac{1}{4} X^2$.

4   If $1 - R_1 \geq W$ exit.

5   If $\exp(-1.35)/R_1 + 0.35 < W$ goto 1.

6   If $-\ln R_1 < W$ goto 1.

7   Exit.

ations) has been reduced markedly by using the squeeze method. A sample of 10 000 normal deviates required 2788 logarithmic evaluations. At the expense of more programming effort, the acceptance probability may be improved by using a tighter bounding enclosing region. Robertson and Walls (1980) use a trapezium-shaped region in FA05A of the Harwell subroutine library, finding it to be approximately 25% faster than a version with rectangular region.

Before moving on to more complex Normal generators it may be helpful to quote some timings for Algorithms N1–N4. Using FORTRAN on a DEC-20 computer, with the random number function RAN (13 μs) these methods took 80, 99, 105, and 85 μs, respectively. The difference between polar Box–Müller and the others can be expected to increase for a slower random number generator, since the others require a larger number of uniform deviates. Coupled with its shortness, the polar Box–Müller method is unlikely to be a bad choice whenever a reasonable compromise between speed and ease of implementation is required. A FORTRAN 77 function RND appears in Appendix 1.1.

4.2.5   Rectangle–wedge–tail method

We now turn to an examination of more involved generators, which are not so easy to implement as the previous ones. The first of these due to Marsaglia *et al.* (1964) is similar in concept to the fast exponential method devised by the same workers. It is often referred to as the 'fast' method, to distinguish it from another, the 'convenient' devised by Marsaglia and Bray (1964). A folded Normal distribution is expressed as a probability mixture,

$$f_X(x) = \left(\frac{2}{\pi}\right)^{\frac{1}{2}} e^{-\frac{1}{2}x^2} = \sum_{i=1}^{3} a_i g_i(x),$$

where

$$a_1 = 0.1 \sum_{j=1}^{30} f_X(0.1j) = 0.9578,$$

$$a_2 = \int_3^\infty f_X(x) dx = 0.0395,$$

$$a_3 = 1 - a_1 - a_2 = 0.0027,$$

$$g_1(x) = a_1^{-1} \begin{cases} f_X(0.1) & (0 \leqslant x \leqslant 0.1) \\ f_X(0.2) & (0.1 < x \leqslant 0.2) \\ \vdots & \\ f_X(3.0) & (2.9 < x \leqslant 3.0) \\ 0 & \text{(elsewhere)} \end{cases}$$

$$g_2(x) = \begin{cases} (f_X(x) - a_1 g_1(x))/a_2 & (0 \leqslant x \leqslant 3.0) \\ 0 & \text{(elsewhere)} \end{cases}$$

$$g_3(x) = \begin{cases} f_X(x)/a_3 & (x > 3) \\ 0 & \text{(elsewhere)}. \end{cases}$$

$g_1(x)$ (the 'rectangular' part) is piecewise uniform. As in the exponential method most of the probability mixture ($a_1 = 0.96$) is concentrated into this distribution. Sampling is via an efficient look-up table method, based on the generation of a discrete random variate. Sampling from $g_2(x)$ (a series of wedges) is via envelope rejection utilizing the approximate piecewise linearity of the function. Sampling from $g_3(x)$ (the 'tail') does not have to be particularly efficient (approximately one in 400 variates is a tail variate). We shall refer to the method used as the 'polar' tail method, to distinguish it from Marsaglia's (1964) 'tail' method. In the polar method two random numbers $R_1, R_2 \sim U(0, 1)$ are taken. Subject to $R_1^2 + R_2^2 \leqslant 1$, set

$$X_1 = [\{9 - 2\ln(R_1^2 + R_2^2)\}/(R_1^2 + R_2^2)]^{\frac{1}{2}} R_1$$

$$X_2 = [\{9 - 2\ln(R_1^2 + R_2^2)\}/(R_1^2 + R_2^2)]^{\frac{1}{2}} R_2.$$

With probability 0.49, at least one of X_1, X_2 will have a value exceeding 3, and in this case has the required distribution $g_3(\cdot)$.

The method requires storage of a large number of constants, and a fairly lengthy and intricate code. No set-up time is required. It has the virtue of requiring an average of slightly in excess of one uniform deviate per (folded) variate, and few time-consuming function evaluations. A FORTRAN implementation on an IBM 360/65 using a slow random number generator (114 μs) predictably resulted in the method being fastest of six exact Normal generators (Atkinson and Pearce, 1976). Atkinson and Pearce found it to be second fastest to the 'convenient' (Section 4.2.6) on a Cyber 73–14 and Univac

1108, where uniform generation took 13 and 12.5 μs, respectively. There is some evidence, therefore, to suggest that the relative speed of the method is fairly robust with respect to a spectrum of computing environments.

4.2.6   The 'convenient' and similar methods

In this section we consider three methods which have some features in common. These are: (i) use of the composition method; (ii) no storage of constants; (iii) a greater requirement for uniform deviates than the 'fast' method; (iv) less coding than the 'fast' method; and (v) no set-up time. Despite (iv) these methods still appear to involve more complex coding than most of the methods given in Sections 4.2.1–4.2.4 and are therefore less easy to implement. The best known of this group is the 'convenient' method of Marsaglia and Bray (1964). The Normal distribution is represented as a probability mixture,

$$f_X(x) = (2\pi)^{-\frac{1}{2}} e^{-\frac{1}{2}x^2} = \sum_{i=1}^{4} a_i g_i(x),$$

where

$g_1(x) \equiv$ p.d.f. of $2(R_1 + R_2 + R_3 - 1.5)$ where
$\qquad R_1, R_2,$ and R_3 are random numbers,

$g_2(x) \equiv$ p.d.f. of $1.5(R_1 + R_2 - 1)$,

$$g_3(x) = \begin{cases} \{f_X(x) - a_1 g_1(x) - a_2 g_2(x)\}/a_3 & (|x| \leqslant 3) \\ 0 & \text{(elsewhere)}, \end{cases}$$

$$g_4(x) = \begin{cases} f_X(x)/a_4 & (|x| > 3) \\ 0 & \text{(elsewhere)}, \end{cases}$$

$a_1 = 0.8638,$
$a_2 = 0.1107,$
$a_3 = 0.0228002039,$
$a_4 = 1 - a_1 - a_2 - a_3 = 0.0026997961.$

Sampling from $g_3(x)$ [that part of $f_X(\cdot)$ unexplained by $g_1(\cdot)$ and $g_2(\cdot)$] uses envelope rejection with a uniform target distribution. Sampling from $g_4(\cdot)$ uses the 'polar' tail method. Note the high dependence on random numbers for the most heavily sampled region of X.

Ahrens and Dieter (1972) use similar ideas in their trapezoidal method (TR). The normal p.d.f. is represented as

$$f_X(x) = \sum_{i=1}^{3} a_i g_i(x),$$

where

$g_1(x)$ has the shape of an isoceles trapezium inscribed in the curve $f_X(x)$
      for $|x| < 2.7140280$,

$g_2(x) = (f_X(x) - a_1 g_1(x))/a_2$ $(|x| < 2.7140280)$,

$g_3(x) = f_X(x)/a_3$ $(|x| \geq 2.7140280)$,

$a_1 = 0.9195444$,

$a_2 = 0.0459427$,

$a_3 = 0.0345129$.

Sampling from $g_1(x)$ is via a linear combination of two random numbers, from $g_3(x)$ via the tail generation procedure of Marsaglia, and from $g_2(x)$ via an envelope rejection procedure.

Kinderman and Ramage (1976) used a probability mixture method, representing the standard normal p.d.f. as

$$f_X(x) = (2\pi)^{-\frac{1}{2}} e^{-\frac{1}{2}x^2} = \sum_{i=1}^{3} a_i g_i(x),$$

where

$g_1(x) = (2.216 - |x|)/(2.216)^2$ $(|x| < 2.216)$,

$g_2(x) = (f_X(x) - a_1 g_1(x))/a_2$ $(|x| < 2.216)$,

$g_3(x) = f_X(x)/a_3$ $(|x| \leq 2.216)$,

$a_1 = 0.884$,

$a_2 = 0.089$,

$a_3 = 0.027$.

The inversion method is used to sample from $g_1(x)$, the polar tail method from $g_3(x)$, and an envelope rejection method comprising a series of triangular target distributions from $g_2(x)$.

Kinderman and Ramage (1976) carried out a comparative empirical study of these three and other methods. Interestingly the three were of very similar speed (274, 279, and 294 μs, respectively, on an IBM 360/67; 82, 80, and 93 μs, respectively, on a SRU 1108). This perhaps reflects the common features and, in particular, a not dissimilar requirement for random numbers (2.92, 2.17, and 2.16, respectively). All three were somewhat faster than the polar Box–Müller method. Atkinson and Pearce (1976) found the 'convenient' method to be faster than both the polar Box–Müller and the 'fast' method, except when using a slow uniform generator on an IBM 360/65. The latter emphasizes that the three methods are not quite so suitable under such conditions. It is worth noting from the many empirical studies performed, that both the polar Box–Müller method and the 'fast' method are rarely much worse than the best of the algorithms tested. Coupled with its simplicity this puts the polar Box–Müller method in a rather favourable position compared with the 'fast' method.

4.2.7   Polynomial sampling and Forsythe's methods

Ahrens and Dieter (1972) have used polynomial sampling in an algorithm
they refer to as TS. They represent the folded normal p.d.f. as

$$f_X(x) = \left(\frac{2}{\pi}\right)^{\frac{1}{2}} e^{-\frac{1}{2}x^2} = \sum_{i=1}^{5} a_i g_i(x),$$

where

$$a_1 = \tfrac{1}{2}, a_2 = \tfrac{1}{4}, a_3 = \tfrac{1}{8}, a_4 = \tfrac{1}{16}, a_5 = \tfrac{1}{16},$$

$$g_1(x) = \begin{cases} f_X(x)/a_1 & (0 < x \leqslant 0.67449) \\ 0 & (\text{elsewhere}) \end{cases}$$

$$g_2(x) = \begin{cases} f_X(x)/a_2 & (0.67449 < x \leqslant 1.1503) \\ 0 & (\text{elsewhere}) \end{cases}$$

$$g_3(x) = \begin{cases} f_X(x)/a_3 & (1.1503 < x \leqslant 1.5341) \\ 0 & (\text{elsewhere}), \end{cases}$$

$$g_4(x) = \begin{cases} f_X(x)/a_4 & (1.5341 < x \leqslant 1.8627) \\ 0 & (\text{elsewhere}) \end{cases}$$

$$g_5(x) = \begin{cases} f_X(x)/a_5 & (x > 1.8627) \\ 0 & (\text{elsewhere}). \end{cases}$$

The p.d.f.s $g_i(x)$ $(i = 1, \ldots, 4)$ are approximated by Taylor series expansions,

$$g_i(x) = \sum_{j=0}^{k} b_j^{(i)} x^j, \qquad (4.9)$$

where k is of the order of 20. The degree of the polynomial ensures that the
approximation is extremely good. Sampling from (4.9) uses the polynomial
sampling method of Section 3.9. Although k is of the order of 20, only on a
very small proportion of occasions will anything approaching k random
numbers be required, since $b_j^{(i)}$ becomes small as j increases. Sampling from
$g_5(x)$ uses Marsaglia's 'tail' method. Ahrens and Dieter (1972) also use
polynomial sampling in algorithm RT, to modify the rectangle–wedge–tail
algorithm. They found FORTRAN implementations of TS and RT on an
IBM 360/50 (random number generation time, 105 μs) took 707 and 647 μs,
respectively, compared with 803 μs for the polar Box–Müller method. Given
the high storage requirement for $\{b_j^{(i)}\}$ and the effort in programming these,
the timings are disappointing and indicate again the respectable performance
of the easily implemented polar Box–Müller Method.

A number of algorithms have arisen from Forsythe's general method.
Using the notation of Section 3.7, the main ones are:

1. Forsythe (1972). Intervals are defined by:

$$q_0 = 0, \, q_j = (2j-1)^{\frac{1}{2}} \qquad (j \geqslant 1),$$

each variate requiring a mean of 3.036 random numbers (excluding the one used for interval selection).

2. Parker (1976) has suggested intervals defined by:

$$q_0 = 0, \, q_j = (2j)^{\frac{1}{2}} \qquad (j \geqslant 1).$$

3. Dieter and Ahrens (1973) in their 'centre-tail' method use one interval defined by:

$$q_0 = 0, \, q_1 = \sqrt{2}.$$

Sampling from the tail $(\sqrt{2}, \infty)$ uses Marsaglia's tail procedure. They claim a mean requirement of 4.879 random numbers.

4. Ahrens and Dieter (1973) (Algorithm FT) use intervals defined by:

$$q_j = R^{-1}(2^j) \qquad (j \geqslant 0),$$

where

$$R(x) = \int_x^\infty \left(\frac{2}{\pi}\right)^{\frac{1}{2}} e^{-\frac{1}{2}u^2} \, du.$$

The mean requirement is 2.539 random numbers, excluding one used for interval selection.

5. Ahrens and Dieter (1973) (Algorithm FL) use intervals defined by:

$$q_0 = 0, \, q_1 = R^{-1}(31/32), \, q_2 = R^{-1}(30/32), \ldots, q_{31} = R^{-1}(1/32),$$
$$q_{32} = R^{-1}(1/64), \, q_{33} = R^{-1}(1/128) \ldots.$$

Because the intervals are of small width, the mean requirement of random numbers would normally be close to 2. Referring to Theorem 3.2, this figure is reduced further by comparing R_1, not with $h_j(X_j)$, but with $h_j^* = \max[h_j(X_j)]$. If $R_1 > h_j^*$ (as it is on most occasions) then $X_j \sim U(q_{j-1}, q_j)$ is delivered. In this case X_j may be generated by 'conditioning' R_1, giving

$$X_j = q_{j-1} + \left(\frac{R_1 - h_j^*}{1 - h_j^*}\right)(q_j - q_{j-1}),$$

without recourse to a separate random number. In this way the mean requirement is reduced to 1.232 random numbers.

6. Brent (1974) has devised a method which combines the intervals defined in (4) with the 'conditioning' technique of (5), giving a mean requirement of 1.37 random numbers.

For the normal distribution, Forsythe-type methods require either a method of generation from the tail of a normal, or truncation of the distribution at a point where the tail probability may be neglected. Atkinson and Pearce (1976) tested types (1) and (6) and found both to be slower than the 'convenient' and 'fast' methods on each of three machines.

4.2.8   Conclusions and recommendations

Amongst the more complex algorithms, the 'fast' and 'convenient' methods perform well under many conditions, the former being particularly suitable when random number generation is slow, and the latter when fast. Regarding easily implemented algorithms, the polar Box–Müller method offers good all-round performance, although the ratio and Butcher's methods are of comparable performance when used with a fast random number generator. For FORTRAN implementations, computational evidence suggests that the best of the easily implemented methods are not substantially slower than the best of the more complex methods, on any given machine.

4.3   Gamma distribution

The gamma distribution with shape parameter α (>0) and scale parameter λ^{-1} ($\lambda > 0$) has p.d.f.

$$f_Z(x) = \frac{\lambda^{\alpha} x^{\alpha-1} e^{-\lambda x}}{\Gamma(\alpha)} \qquad (x \geqslant 0).$$

Variates may be generated by sampling from a standard gamma distribution having p.d.f.

$$f_X(x) = \frac{x^{\alpha-1} e^{-x}}{\Gamma(\alpha)} \qquad (x \geqslant 0),$$

and setting $Z = X/\lambda$.

When α is integer valued, one method which appears to be frequently used, but is not recommended, is the practice of delivering X as the sum of α independent standard exponential variates. This gives.

$$X = - \sum_{i=1}^{\alpha} \ln R_i.$$

To avoid a possible large number of logarithmic evaluations, this can be rewritten as

$$X = -\ln \prod_{i=1}^{\alpha} R_i.$$

There are several disadvantages to this method. Firstly, it is applicable only for integral values of α. Secondly, for moderate or large α, it becomes extremely slow due to the requirement for α random numbers. Finally, the logarithmic evaluation will overflow if the product of the random numbers is less than the smallest representable real number on the computer. Taking a machine with a hypothetical floating point underflow at 10^{-37}, the procedure will attempt to compute the logarithm of zero if the true value exceeds $-\ln (10^{-37}) = 85.2$. Of course a variate of this magnitude is likely to arise only for large values of α, when the method would be inappropriate in any case.

4.3.1 A ratio method ($\alpha > 1$)

In Section 3.6 we showed how a standard application of the ratio of uniforms approach gave an asymptotic sampling efficiency of $\sqrt{\{\pi/(2\alpha)\}}$, indicating that the method is not robust for large α. We noted, however, that a modified approach due to Kinderman and Monahan (1980), in which the original distribution is relocated so that the mode is now at zero, gives an asymptotic sampling efficiency of $\sqrt{(\pi e)}/4$. Furthermore, the acceptance region plots in Fig. 3.5 indicate that the sampling efficiency is of a uniformly high value for all values of α. Using the results of Section 3.6, we derive Algorithm G1.

When α is not reset between calls of the generator, values of a and b may be saved. X' is a relocated variate, while X is a variate from the original distribution. In step 4, a sufficient condition for rejection is that $X < 0$. Step 7 is the rejection test (see result 3.12). Step 6 is an acceptance pretest based on the inequality $\ln\{X/(\alpha-1)\} \geqslant 1 - \{(\alpha-1)/X\}$. Note that at most two logarithmic evaluations per prospective variate are required.

4.3.2 Cauchy method ($\alpha \geqslant 1$)

Ahrens and Dieter (1974) have proposed an algorithm (GC) using envelope rejection with a Cauchy target distribution,

$$h_Y(x) = b/[\pi\{b^2 + (x-a)^2\}],$$

Algorithm G1

Input $[\alpha]$ Saved $[a, b]$
Output $[X]$

S1 $a := \{1 + \sqrt{(2\alpha-1)}\}\left\{\dfrac{\alpha + \sqrt{(2\alpha-1)}}{\alpha-1}\right\}^{(\alpha-1)/2} e^{-\{1+\sqrt{(2\alpha-1)}\}/2}.$

S2 $b := 2(1-\alpha)/(ea)$.
1   generate $R_1, R_2 \sim U(0, 1)$.
2     $X' := a(R_2/R_1) + b\{(1-R_2)/R_1\}$.
3     $X := X' + \alpha - 1$.
4   If $X < 0$ goto 1.
5     $V := X' + 2 \ln R_1$.
6   If $(\alpha-1)X'/X \geqslant V$ exit.
7   If $(\alpha-1) \ln \{X/(\alpha-1)\} < V$ goto 1.
8   Exit.

where a and b are constants. Since negative values of Y have to be discarded, it is better to use a truncated distribution,

$$g_Y(x) = Ah_Y(x) \qquad (x \geqslant 0), \qquad (4.10)$$

where $A^{-1} = \{\pi + 2\tan^{-1}(a/b)\}/(2\pi)$. This gives

$$\frac{f_X(x)}{g_Y(x)} = \frac{x^{\alpha-1}e^{-x}\pi\{b^2 + (x-a)^2\}}{Ab\Gamma(\alpha)}, \qquad (4.11)$$

which has turning points at

$$(\alpha - 1 - x)\{b^2 + (x-a)^2\} + 2x(x-a) = 0.$$

To facilitate root finding, a and b are selected as

$$a = \alpha - 1, \; b = \sqrt{(2\alpha - 1)},$$

giving $(\alpha - 1 - x)(x - \alpha)^2 = 0$. For $\alpha \geqslant 1$, it turns out that $x = \alpha - 1$ is a maximum, while $x = \alpha$ is a point of inflection. Thus

$$M = \max[f_X(x)/g_Y(x)]$$
$$= (\alpha - 1)^{\alpha-1}e^{-(\alpha-1)}\pi\sqrt{(2\alpha - 1)}/\{A\Gamma(\alpha)\}. \qquad (4.12)$$

When $\alpha = 1$, $M = \pi/2$. As α increases, M becomes slightly smaller, then rises to its asymptotic value (obtainable using Stirling's formula) of $\sqrt{\pi} = 1.77$. M is therefore bounded and the sampling efficiency reasonably high for all values of α, a necessary requirement for a robust procedure.

By forming the c.d.f. $G_Y(x)$ from eqn (4.10), inversion can be used to generate prospective variates Y according to

$$Y = a + b\tan[R_1\{(\pi/2) + \tan^{-1}(a/b)\} - \tan^{-1}(a/b)].$$

Given a second random number R_2, eqns (4.11) and (4.12) lead to the acceptance condition

$$R_2 < \{Y/(\alpha - 1)\}^{\alpha-1}e^{-(Y-\alpha+1)}\left\{1 + \left(\frac{Y-a}{b}\right)^2\right\},$$

or

$$\ln\{R_2/(1 + W^2)\} < (\alpha - 1)\ln\{Y/(\alpha - 1)\} - Y + \alpha - 1,$$

where $W = (Y - a)/b$. Details of a suitable procedure are shown in Algorithm G2.

Note that the set-up time is dominated by a square root and inverse tangent evaluation, while each prospective variate requires one tangent and two logarithmic evaluations. The latter can be reduced through the use of suitable pretests.

Algorithm G2

Input $[\alpha]$ Saved $[a, b, c, d]$
Output $[X]$

S1 $a := \alpha - 1$, $b := \sqrt{(2\alpha - 1)}$, $c := \tan^{-1}(a/b)$, $d := (\pi/2) + c$.
1   generate $R_1, R_2 \sim U(0, 1)$.
2    $W := \tan(dR_1 - c)$, $X := a + bW$.
3   If $\ln\{R_2/(1 + W^2)\} \geqslant a\ln(X/a) - X + a$ goto 1.
4   Exit.

4.3.3   Log–logistic method ($\alpha > 0$)

Another envelope rejection method has been proposed by Cheng (1977). It uses a log–logistic target distribution,

$$g_Y(x) = \lambda\mu x^{\lambda - 1}/(\mu + x^\lambda)^2 \qquad (x \geqslant 0), \tag{4.13}$$

where $\mu = \alpha^\lambda$, $\lambda = \sqrt{(2\alpha - 1)}$ for $\alpha \geqslant 1$, and $\lambda = \alpha$ for $\alpha < 1$. With these parameter values,

$$\frac{f_X(x)}{g_Y(x)} = \frac{(\mu + x^\lambda)^2 \, x^{\alpha - \lambda}e^{-x}}{\lambda\mu\Gamma(\alpha)},$$

which has a global maximum at $x = \alpha$. This gives

$$M = \max[f_X(x)/g_Y(x)]$$

$$= \frac{4\alpha^\alpha e^{-\alpha}}{\Gamma(\alpha)\sqrt{(2\alpha - 1)}}.$$

When $\alpha = 1$, $M = 4/e = 1.471$, decreasing to an asymptotic value of $2/\sqrt{\pi}$ $= 1.128$. M still remains bounded as α decreases below 1, reaching a maximum of 4 as $\alpha \to 0$. The method is therefore exceptional in having a bounded execution time for all values of the parameter α. However, the procedure is recommended only when $\alpha \geqslant 1$, where the sampling efficiency is *consistently* high.

Using the inversion method on (4.13),

$$Y = \alpha\{R_1/(1 - R_1)\}^{1/\lambda}, \tag{4.14}$$

or equivalently $Y = \alpha e^V$, where $V = \ln\{R_1/(1 - R_1)\}/\sqrt{(2\alpha - 1)}$. Given a second random number the acceptance condition is

$$R_2 < \tfrac{1}{4}\{1 + (Y/\alpha)^\lambda\}^2 (Y/\alpha)^{\alpha + \lambda}e^{-(Y - \alpha)}/(Y/\alpha)^{2\lambda}.$$

Using eqn (4.14) this becomes

$$R_1^2 R_2 < \tfrac{1}{4}(Y/\alpha)^{\alpha + \lambda}e^{-(Y - \alpha)}$$

or

$$\ln(R_1^2 R_2) < -\ln 4 + (\alpha + \lambda)\ln(Y/\alpha) - Y + \alpha.$$

This results in Algorithm G3, which is shown for $\alpha \geqslant 1$. Step 7 is the rejection condition. Step 6 is an acceptance pretest based on the inequality $\ln(R_1^2 R_2) \leqslant \theta R_1^2 R_2 - 1 - \ln \theta$ for any $\theta > 0$. The value of θ is not critical and 4.5 has been suggested.

Algorithm G3

Input $[\alpha]$ Saved $[a, b, c]$
Output $[X]$ Parameter $[\theta = 4.5]$.

S1 $a := (2\alpha - 1)^{-\frac{1}{2}}, b := \alpha - \ln 4, c := \alpha + 1/a.$
1   generate $R_1, R_2 \sim U(0, 1).$
2     $V := a \ln\{R_1/(1 - R_1)\}.$
3     $X := \alpha e^V.$
4     $Z := R_1^2 R_2.$
5     $T := b + cV - X.$
6   If $\theta Z \leqslant 1 + \ln \theta + T$ exit.
7   If $\ln Z > T$ goto 1.
8   Exit.

4.3.4   *t*-distribution method ($\alpha > 1$)

An ingenious method which has the merit of a very low set-up time is due to Best (1978). Envelope rejection is again used, generating a prospective variate $Y = \alpha - 1 + \beta T$, where T is a Student's t-variate with two degrees of freedom. The term $(\alpha - 1)$ relocates the distribution so that the mode is identical to that of the gamma, while β is a suitable scaling constant.

The unusual choice of target distribution is presumably motivated by the fact that T may be obtained using inversion. Since

$$f_T(x) = \frac{\{1 + (x^2/2)\}^{-\frac{3}{2}}}{2\sqrt{2}}, \tag{4.15}$$

$$F_T(x) = \tfrac{1}{2}\{1 + x/\sqrt{(x^2 + 2)}\}.$$

Given a random number R_1, inversion yields

$$R_1 = \tfrac{1}{2}\{1 + T/\sqrt{(T^2 + 2)}\}$$

or

$$(2R_1 - 1)^2 = T^2/(T^2 + 2).$$

Thus

$$T^2 = \frac{2(R_1 - \tfrac{1}{2})^2}{R_1(1 - R_1)}.$$

Since T is symmetrically distributed, roots

$$\pm \left\{ \frac{2(R_1 - \tfrac{1}{2})^2}{R_1(1 - R_1)} \right\}^{\tfrac{1}{2}}$$

are chosen with equal probability, a trivial instance of transformations involving multiple roots—see Section 3.10. This is most easily done by setting

$$T = \frac{2^{\tfrac{1}{2}}(R_1 - \tfrac{1}{2})}{\{R_1(1 - R_1)\}^{\tfrac{1}{2}}} \tag{4.16}$$

From eqn (4.15), the p.d.f. of Y is derived as

$$g_Y(y) = \left\{ 1 + \tfrac{1}{2}\left(\frac{y - \alpha + 1}{\beta}\right)^2 \right\}^{-\tfrac{3}{2}} \bigg/ (2\sqrt{2}\beta).$$

Thus

$$\frac{f_X(x)}{g_Y(x)} = \frac{\beta x^{\alpha - 1} e^{-x} 2\sqrt{2} \left\{ 1 + \tfrac{1}{2}\left(\frac{x - \alpha + 1}{\beta}\right)^2 \right\}^{\tfrac{3}{2}}}{\Gamma(\alpha)},$$

which has a maximum at $x = \alpha - 1$ and a point of inflection at $x = \alpha + 0.5$ when the scaling constant is chosen as

$$\beta = \{(3\alpha - 0.75)/2\}^{\tfrac{1}{2}}.$$

Thus

$$M = f_X(\alpha - 1)/g_Y(\alpha - 1)$$
$$= 2\sqrt{(3\alpha - 0.75)}(\alpha - 1)^{\alpha - 1} e^{-(\alpha - 1)}/\Gamma(\alpha). \tag{4.17}$$

The acceptance condition follows immediately as

$$R_2 < \{ Y/(\alpha - 1) \}^{\alpha - 1} e^{-Y + \alpha - 1} \{ 1 + \tfrac{1}{2}T^2 \}^{\tfrac{3}{2}}.$$

Using eqn (4.16), this becomes

$$R_2 < \{ Y/(\alpha - 1) \}^{\alpha - 1} e^{-Y + \alpha - 1} [4(R_1 - R_1^2)]^{-\tfrac{3}{2}}$$

or

$$\ln d < 2[(\alpha - 1)\ln\{ Y/(\alpha - 1) \} - f],$$

where

$$d = 64 R_2^2 (R_1 - R_1^2)^3,$$

and

$$f = \beta T = \left[\frac{3\alpha - 0.75}{R_1 - R_1^2} \right]^{\tfrac{1}{2}} (R_1 - \tfrac{1}{2}).$$

Algorithm G4 shows the details.

The major advantage of this procedure is that virtually no time is required to set-up the constant values b and c in step S1. It is thus particularly suitable when α changes from one call to another. Equation (4.17) indicates that the

Algorithm G4

Input [α]   Saved [b, c]
Output [X]

S1 $b := \alpha - 1$, $c := 3\alpha - 0.75$.
1   generate $R_1 \sim U(0, 1)$.
2    $g := R_1 - R_1^2$.
3    $f := (R_1 - \frac{1}{2})(c/g)^{\frac{1}{2}}$.
4    $X := b + f$.
5   If $X \leqslant 0$ goto 1.
6   generate $R_2 \sim U(0, 1)$.
7    $d := 64R_2^2 g^3$.
8   If $d < (1 - 2f^2/X)$ exit.
9   If $\ln d \geqslant 2\{b \ln(X/b) - f\}$ goto 1.
10   Exit.

sampling efficiency is adequately high for most values of α, although $M \rightarrow 3$ as $\alpha \rightarrow 1$. The latter is not as bad as might appear, since a large proportion of rejected variates will be excluded at step 5. As α increases, M decreases fairly rapidly taking values 1.874, 1.685, 1.475 at $\alpha = 1.5$, 2, and 5, respectively, approaching its asymptotic value of $(6/\pi)^{\frac{1}{2}} = 1.382$, as $\alpha \rightarrow \infty$. Marginal generation time is further reduced through the use of an acceptance pretest in step 8.

4.3.5   Comparison of algorithms suitable for $\alpha \geqslant 1$

At this point it will be useful to introduce some empirical evidence to complement the theoretical analyses of Algorithms G1–G4. The experimental evidence offered is not intended to be exhaustive, since it relates only to performance on one machine with one type of random number generator. It is sufficient, however, to differentiate any gross differences in performance, which are quite likely to occur whatever the computing environment.

The four algorithms were programmed as FORTRAN subroutines and run on a DEC-20, using the random number function RAN, referred to in Sections 2.4.1 and 2.4.8. Each uniform deviate takes approximately 13 μs to generate. The mean execution time for each generator is obtained by creating samples of size 1000 and 11 000 and differencing the times. Timing experiments were conducted when demand by other users was low, so as to reduce variation due to multi-programming. Under these conditions, the maximum variation appears to be of the order of 7%. Although this is still quite high, it is not too large to mask any gross differences, and it is only these ones which are of relevance when extrapolating conclusions to a wider computing environment. Two sets of experiments were carried out, one when α was reset between calls, and the other in which it remained fixed between such calls. The former, of course, carry an overhead of set-up calculations, while in the latter, parameter values may be saved. Table 4.1 shows the mean execution times.

Table 4.1 Mean time (μs) to generate one gamma variate (FORTRAN implementation) on a DEC-20 computer

| Method | α | | | | |
	$1 + \varepsilon^\dagger$	2	5	10	100
Ratio (G1)	191 (410)	206 (433)	235 (458)	251 (472)	276 (500)
Cauchy (G2)	336 (397)	287 (370)	311 (394)	330 (412)	360 (444)
Log–logistic (G3)	296 (329)	237 (285)	221 (263)	217 (262)	211 (263)
t-distribution (G4)	362 (364)	231 (241)	232 (235)	229 (240)	232 (238)

† $\varepsilon = 10^{-4}$ for G1 and G4, and zero for G2 and G3.
Numbers in parentheses are timings when α is reset between calls.

The timings reflect the behaviour of sampling efficiency in each case. The ratio method gives good overall performance when α is fixed between calls, but is clearly inappropriate when α is reset. In contrast, the t-distribution method requires virtually no set-up time, yet gives competitive performance for most values of α. In view of this, the method is recommended when $\alpha \geqslant 1$, and a FORTRAN 77 function GAMMA1 appears in Appendix 1.3.

We have concentrated mainly on those methods that are short, easy to code, require few set-up calculations, and give a uniformly high level of performance for all $\alpha \in (1, \infty)$. There are many other procedures available, some of which do not satisfy these criteria. Ahrens and Dieter (1974) in their algorithm GO use a target distribution which is a mixture of exponential and normal. It is suitable only when $\alpha > 2.53$. Wallace (1974) also uses envelope rejection with a target distribution which is a mixture of Erlang distributions, having shape parameters $\langle \alpha \rangle$ and $\langle \alpha + 1 \rangle$, respectively. The current state of the art gives no advantage to such a method, since it is now just as easy to sample from a gamma (α non-integer) as it is from an Erlang distribution. Fishman (1976a) uses an exponential target distribution, which provides a poor fit as α increases. Greenwood (1974) exploits the Wilson–Hilferty (1931) approximation to the chi-squared in a method suitable when $\alpha > \frac{1}{2}$. However, a small numerical optimization is required. Atkinson (1977b) developed a 'switching' algorithm, with a target distribution which is a mixture of uniform and exponential. The performance is comparable to the log–logistic method for $\alpha \lesssim 4$, but since the asymptotic sampling efficiency is $O(\alpha^{-1/2})$, it is not competitive for larger values. Tadikamalla (1978a, b) proposes two envelope methods, with Erlang and Laplace target distributions, respectively.

Schmeiser and Lal (1980) develop more complex methods including one where the target distribution is a mixture of six uniform, two triangular, and two exponential distributions. These methods require increased set-up time and memory requirements over other methods. Cheng and Feast (1979) used a ratio of uniforms method with a parallelogram as an enclosing region. This provides a high sampling efficiency for all $\alpha \geqslant 1$. Finally Atkinson and Pearce (1976) report on a Forsythe implementation, with presumably some truncation of the distribution.

4.3.6 A power transformation method ($\alpha \leqslant 1$)

We now turn to situations where the shape parameter does not exceed 1. This requires different methods from those given previously, due to the unbounded nature of the p.d.f. as $x \to 0$ when $\alpha < 1$. One approach is to find a suitable transformation of X. Following the suggestion made in Section 3.6, we consider $Z = X^{\alpha}$, and the reciprocal variate $W = Z^{-1}$. These have p.d.f.s,

$$f_Z(z) = e^{-z^{1/\alpha}}/\Gamma(\alpha+1) \qquad (z \geqslant 0),$$
$$f_W(w) = e^{-w^{-1/\alpha}}/\{w^2\Gamma(\alpha+1)\} \qquad (w \geqslant 0).$$

These two functions are maximized (for all α) at $m_Z = 0$ and $m_W = (2\alpha)^{-\alpha}$, respectively. Thus the acceptance region in a ratio implementation is enclosed by a rectangle with dimensions

$$u_+ = f_Z^{\frac{1}{2}}(m_Z) = \{\Gamma(\alpha+1)\}^{-\frac{1}{2}}$$

$$v_+ = f_W^{\frac{1}{2}}(m_W) = \left(\frac{2\alpha}{e}\right)^{\alpha}\{\Gamma(\alpha+1)\}^{-\frac{1}{2}}.$$

A prospective (transformed) variate is therefore

$$Z = \frac{R_2 v_+}{R_1 u_+} = \left(\frac{2\alpha}{e}\right)^{\alpha}\frac{R_2}{R_1}$$

or

$$X = Z^{1/\alpha} = \left(\frac{2\alpha}{e}\right)\left(\frac{R_2}{R_1}\right)^{1/\alpha}$$

The acceptance condition is

$$R_1 u_+ < f_Z^{\frac{1}{2}}(Z)$$

or

$$R_1 < e^{-0.5Z^{1/\alpha}}.$$

Equivalently this is

$$-2\ln R_1 > Z^{1/\alpha} = X.$$

This gives rise to an easily implemented algorithm, G5. For moderately small values of α ($\lesssim 0.2$) X can underflow. Step 2 anticipates this, setting $X := 0$,

Algorithm G5

Input $[\alpha]$ Saved $[a, b, c, d]$
Output $[X]$ Parameter $[U, V$ = machine's smallest and largest positive
                real numbers.$]$

S1 $a := \left(\dfrac{2\alpha}{e}\right), b := \dfrac{1}{\alpha}, c := \left(\dfrac{2U}{a}\right)^{\alpha}, d := V^{\alpha}.$

1   generate $R_1, R_2 \sim U(0, 1)$.
2   If $R_2 \leqslant R_1 c$ then
3        $X := 0$
4   Else If
5        $R_2 > R_1 d$ then
6   goto 1
7   Else
8        $X := a \left(\dfrac{R_2}{R_1}\right)^{b}$
9        $W := \frac{1}{2}X$
10   If $1 - R_1 > W$ exit
11   If $1/R_1 - 1 \leqslant W$ goto 1
12   If $-\ln R_1 \leqslant W$ goto 1
13   End If.
14   Exit.

thereby speeding up delivery of variates. Similarly step 5 anticipates whether overflow would occur. If so, rejection follows, as condition 11 would always be satisfied for very large X. The acceptance probability for this method is

$$P = \frac{0.5}{u_{+}v_{+}} = \frac{1}{2}\Gamma(\alpha + 1) \left(\frac{e}{2\alpha}\right)^{\alpha}$$

which rises from 0.5 ($\alpha = 0$) to a maximum of 0.731 ($\alpha \cong 0.54$) decreasing to $e/4 = 0.680$ ($\alpha = 1$).

4.3.7   A switching algorithm ($\alpha < 1$)

In this approach envelope rejection is used with a target distribution

$$g_Y(x) = \begin{cases} p\alpha x^{\alpha - 1}/t^{\alpha} & (0 \leqslant x \leqslant t) \\ (1 - p)e^{-(x - t)} & (x > t), \end{cases} \tag{4.18}$$

where $0 \leqslant p \leqslant 1$ and t are free parameters. The form of $g_Y(\cdot)$ ensures that $f_X(x)/g_Y(x)$ is bounded as $x \to 0$. The term 'switching algorithm' appears first to have been used by Atkinson and Whittaker (1976) in connection with beta variate generation. It describes a rejection/composition method where the target distribution is a probability mixture, in which the component p.d.f.s cover mutually exclusive regions of $g_Y(\cdot)$'s domain. Thus, in eqn (4.18) a density $\alpha x^{\alpha-1}/t^\alpha$ $(x \leqslant t)$ is 'switched' in with probability p, while $e^{-(x-t)}$ $(x > t)$ is 'switched' with probability $(1-p)$. Using inversion on eqn (4.18), prospective variates are generated via

$$Y = \begin{cases} t(R_1/p)^{1/\alpha} & (R_1 \leqslant p) \\ t + \ln\left(\dfrac{1-p}{1-R_1}\right) & (R_1 > p). \end{cases}$$

An optimal choice of t and p will be those which minimize $M = \max[\,f_X(x)/g_Y(x)\,]$. From eqn (4.18)

$$\frac{f_X(x)}{g_Y(x)} = \begin{cases} e^{-x}t^\alpha/\{p\Gamma(\alpha+1)\} & (0 \leqslant x \leqslant t) \\ x^{\alpha-1}e^{-t}/\{(1-p)\Gamma(\alpha)\} & (x > t). \end{cases} \qquad (4.19)$$

For fixed t, M is minimized by ensuring that the maximum of $f_X(x)/g_Y(x)$ in $[0, t)$ is identical to its maximum in (t, ∞). Since these maxima occur at 0 and t_+, respectively, the optimized value of p is determined by

$$\frac{t^\alpha}{p\Gamma(\alpha+1)} = \frac{t^{\alpha-1}e^{-t}}{(1-p)\Gamma(\alpha)}$$

or

$$p = t/(t + \alpha e^{-t}).$$

This gives

$$M = f_X(0)/g_Y(0)$$
$$= \frac{t^{\alpha-1}(t + \alpha e^{-t})}{\Gamma(\alpha+1)}.$$

Minimizing M w.r.t. t gives

$$t(e^t - 1) = 1 - \alpha.$$

To avoid this numerical optimization, we choose $t = 1 - \alpha$. This results in only a small loss in efficiency as shown in Table 4.2. The method is of course exact, regardless of the value of t used.

Table 4.2 Sampling efficiencies for switching algorithm using optimized and non-optimized t

t	α							
	0.01	0.1	0.3	0.5	0.7	0.8	0.9	1.0
Optimized	0.991	0.921	0.824	0.785	0.798	0.828	0.882	1.000
Non-optimized	0.991	0.920	0.824	0.780	0.774	0.790	0.836	1.000

On using eqn (4.19), the acceptance condition becomes

$$R_2 < \begin{cases} e^{-Y} & (0 \leqslant Y \leqslant t) \\ (Y/t)^{\alpha-1} & (Y > t). \end{cases}$$

The full method is shown in Algorithm G6. Steps 10 and 11 are pretests. Step 2 checks for underflow of $t(R_1/p)^c$, returning $X = 0$ when this occurs. Finally we note that Ahrens and Dieter (1974) have devised a similar algorithm named GS in which $t = 1$. This gives a uniformly lower sampling efficiency (e.g. 0.723 and 0.731 at $\alpha = 0.9$ and 1.0, respectively).

Algorithm G6

Input [α]   Saved [t, p, c, d]
Output [X]   Parameter [U = machine's smallest positive real number]

S1 $t := 1 - \alpha$, $p := t/(t + \alpha e^{-t})$, $c := 1/\alpha$, $d := p(U/t)^\alpha$.
1   generate $R_1 \sim U(0, 1)$.
2   If $R_1 \leqslant d$ then
3        $X := 0$, Exit
4   Else If $R_1 \leqslant p$ then
5        $X := t\left(\dfrac{R_1}{p}\right)^c$, $W := X$
6   Else
7        $X := t + \ln\{(1-p)/(1-R_1)\}$, $W := t \ln(X/t)$
8   End If.
9   generate $R_2 \sim U(0, 1)$.
10   If $1 - R_2 > W$ exit.
11   If $1/R_2 - 1 \leqslant W$ goto 1.
12   If $-\ln R_2 \leqslant W$ goto 1.
13   Exit.

4.3.8   Beta method ($\alpha < 1$), and a comparison of methods

Jöhnk (1964) sets $X = EY$, where E is a standard exponential variate and Y is a beta variate (see Section 4.4) with parameters α and $1 - \alpha$. The latter is obtained by delivering

$$Y = R_1^{1/\alpha}/(R_1^{1/\alpha} + R_2^{1/(1-\alpha)}),$$

subject to $R_1^{1/\alpha} + R_2^{1/(1-\alpha)} \leqslant 1$, where R_1 and $R_2 \sim U(0, 1)$. The probability of acceptance never falls below $0.25\pi = 0.785$, although the generation of one exponential variate, and two exponential evaluations on each trial can be quite time consuming. The method is, however, easily implemented as Algorithm G7. When programming, the method execution time is reduced by catering for underflow as $\alpha \to 0, 1$.

Algorithm G7

Input [α]   Saved [a, b]
Output [X]

S1 $a := 1/\alpha$, $b := 1/(1-\alpha)$.
1   generate $R_1, R_2 \sim U(0, 1)$.
2    $W := R_1^a$, $V := R_2^b$.
3    $T := W + V$.
4   If $T > 1$ goto 1.
5   generate $R_3 \sim U(0, 1)$.
6    $X := -W \ln R_3/T$.
7   Exit.

At this point we may compare the three methods suitable for $\alpha < 1$. Timings obtained in the manner described in Section 4.3.5 appear in Table 4.3. G5 is suitable when $\alpha = 1.0$. The other two require small amounts of additional code for this special case. The low marginal generation time for G5 as $\alpha \to 0$, results from most of the variates being zero (within machine precision), for which no time-consuming transformation of random numbers is required. Similar effects are evident for G6 (as $\alpha \to 0$) and G7 (as $\alpha \to 0, 1$). The rather high set-up times may be reduced for non-extreme values of α by removing code for testing under or overflow.

Table 4.3 Mean time (μs) to generate one gamma variate (FORTRAN implementations) on a DEC-20 computer

	α					
Method	10^{-4}	0.2	0.4	0.6	0.8	0.999
Power (G5)	85	254	260	263	275	273
	(308)	(512)	(504)	(501)	(521)	(524)
Switching (G6)	42	194	224	210	196	146
	(308)	(451)	(462)	(459)	(478)	(388)
Beta (G7)	43	393	417	426	392	110
	(170)	(537)	(578)	(572)	(534)	(249)

Numbers in parentheses are timings when α is reset between calls.

The timings indicate that the switching method G6 is, with one exception uniformly fastest for α fixed and reset between calls. It can be recommended as a procedure which is robust, easily implemented, and efficient. A FORTRAN 77 function GAMMA2 appears in Appendix 1.4.

The power transformation method is a little disappointing. In part this is due to the time required to obtain the prospective variate in step 8. Cheng and Feast (1980) consider a transformation $Z = X^{1/n}$, where n is an integer with $n \geqslant \alpha^{-1}$. At the expense of accepting a lower limit on α (e.g. 0.25 when $n = 4$) the power transformation may be avoided by evaluating $X = Z^n$ as a product. They reported a useful improvement on the GS algorithm of Ahrens and Dieter, when using a CDC 7600 (RANF uniform generator) and an ICL 2980 (NAG G05AAF generator). For the few applications where α is small, the advantage will not be so great, as n will then be large.

We conclude this section by noting that α may vary during the course of a simulation. In these circumstances it is advisable to use the calling function GAMMA (Appendix 1.2) which selects GAMMA1, GAMMA2, or EXPON according to whether $\alpha > 1$, $\alpha < 1$, or $\alpha = 1$, respectively.

4.4 Beta distribution

A random variable is β-distributed with shape parameters α and β if its p.d.f. is

$$f_{X_{\alpha,\beta}}(x) = \frac{x^{\alpha-1}(1-x)^{\beta-1}}{B(\alpha, \beta)} \qquad (0 \leqslant x \leqslant 1), \tag{4.20}$$

where $\alpha > 0$, $\beta > 0$ and $B(\alpha, \beta)$ is the beta function. Its usefulness derives from the richness of shapes offered and the finite domain of the p.d.f. The methods of generation available frequently reflect the three types of behaviour. These are 'bridge'-shaped ($\alpha > 1$ and $\beta > 1$), 'J'-shaped ($\alpha \leqslant 1$ and $\beta \geqslant 1$ or $\alpha \geqslant 1$ and $\beta \leqslant 1$), and 'U'-shaped ($\alpha < 1$ and $\beta < 1$). If $\alpha < 1$ or $\beta < 1$ the density is unbounded at $x = 0, 1$, respectively. There is great advantage in devising a generator that can cope with all three cases (including the degenerate case $\alpha = \beta = 1$). For this reason one may be prepared to trade-off some simplicity in methods and programming and perhaps even a certain amount of efficiency, against the convenience of not having to switch between three generators, perhaps within one simulation experiment.

We consider in detail four generators, three of which are suitable for all values of the shape parameters. Reference is then made to several other methods, each of which appears to have at least one significant disadvantage.

4.4.1 Gamma method

It is a simple exercise to show that if X_α and X_β are independent standard gamma random variables with shape parameters α and β, then $X_{\alpha,\beta} = X_\alpha/(X_\alpha + X_\beta)$. This gives rise to an easily implemented generator, suitable for all values of α and β, having approximately half the speed of a gamma generator. It is recommended that G4 and G6 be used for parameter values greater/less than 1, respectively.

4.4.2 Jöhnk's rejection method

Given two random numbers R_1 and R_2, Jöhnk (1964) showed that if $R_1^{1/\alpha} + R_2^{1/\beta} \leqslant 1$, then $X_{\alpha,\beta} = R_1^{1/\alpha}/(R_1^{1/\alpha} + R_2^{1/\beta})$. The sampling efficiency is $\Gamma(\alpha+1)\Gamma(\beta+1)/\Gamma(\alpha+\beta+1)$, which shows that this easily implemented procedure is reasonably efficient when α and β are both $\leqslant 1$. In this instance the sampling efficiency lies between 0.5 and 1 (e.g. 0.5 for $\alpha = \beta = 1$, $\pi/4$ for $\alpha = \beta = 0.5$, and 1 as $\alpha,\beta \to 0$). The method is not recommended when one parameter is larger than 1, and the other smaller than 1, as the efficiency decreases when the value of the larger parameter increases (e.g. 0.589 for $\alpha = 0.5$, $\beta = 1.5$ and 0.369 for $\alpha = 0.5$, $\beta = 5$). The method is poor when both parameter values exceed 1, the efficiency decreasing rapidly with increasing parameter values. Although an attractive method when both parameters are less than 1, the two power transformations can make the method slower than others.

4.4.3 Log–logistic method

Cheng (1978) notes that if W has a beta prime distribution with p.d.f.,

$$f_W(x) = x^{\alpha-1}(1+x)^{-(\alpha+\beta)}/B(\alpha,\beta) \qquad (x \geqslant 0), \tag{4.21}$$

then $X_{\alpha,\beta} = W/(1+W)$. The problem of generation therefore reduces to sampling from (4.21). Note that one method is to take the ratio of two gamma variates, $(W = X_\alpha/X_\beta)$, which gives the gamma method. For certain parameter values it is faster to use envelope rejection with a log–logistic target distribution

$$g_Y(x) = \lambda\mu x^{\lambda-1}/(\mu+x^\lambda)^2 \qquad (x \geqslant 0), \tag{4.22}$$

where λ and μ are positive constants. From eqns (4.21) and (4.22)

$$\frac{f_W(x)}{g_Y(x)} = \frac{x^{\alpha-\lambda}(\mu+x^\lambda)^2}{\lambda\mu B(\alpha,\beta)\,(1+x)^{\alpha+\beta}}. \tag{4.23}$$

By choosing $\mu = (\alpha/\beta)^\lambda$, $f_W(x)/g_Y(x)$ conveniently has a turning point at $x = \alpha/\beta$. It may be shown that this is a global maximum if the second free parameter is chosen as

$$\lambda = \begin{cases} \min(\alpha,\beta) & (\min(\alpha,\beta) \leqslant 1) \\ \sqrt{\{(2\alpha\beta - \alpha - \beta)/(\alpha+\beta-2)\}} & (\min(\alpha,\beta) > 1). \end{cases} \tag{4.24}$$

This gives

$$M = f_W(\alpha/\beta)/g_Y(\alpha/\beta)$$

$$= \frac{4\alpha^\alpha\beta^\beta}{\lambda B(\alpha,\beta)\,(\alpha+\beta)^{\alpha+\beta}} \tag{4.25}$$

where λ is as indicated.

Prospective variates may be generated by inversion on (4.22) giving

$$Y = \{\mu R_1/(1 - R_1)\}^{1/\lambda}, \tag{4.26}$$

or equivalently $Y = de^G$, where $G = \lambda^{-1} \ln\{R_1/(1 - R_1)\}$ and $d = \mu^{1/\lambda} = \alpha/\beta$. The acceptance condition, from eqn (4.23) is

$$R_2 \leqslant \frac{1}{4}\left(\frac{Y}{d}\right)^{\alpha-\lambda}\left\{1 + \left(\frac{Y}{d}\right)^{\lambda}\right\}^2 \left(\frac{1+d}{1+Y}\right)^{\alpha+\beta}$$

which reduces to

$$4R_2R_1^2 \leqslant \left(\frac{Y}{d}\right)^{\alpha+\lambda}\left(\frac{1+d}{1+Y}\right)^{\alpha+\beta},$$

on using eqn (4.26).

Equation (4.26) deserves particular attention as overflow will occur if

$$R_1 \geqslant \left\{1 + \left(\frac{\alpha}{\beta V}\right)^{\lambda}\right\}^{-1}, \tag{4.27}$$

where V is the machine's largest representable real number. We may assume that $\beta \leqslant \alpha$. (If $\alpha < \beta$, then deliver $X_{\alpha, \beta} = 1 - X_{\beta, \alpha}$.) If $\min(\alpha, \beta) > 1$, then $\lambda > 1$ and (4.27) will be satisfied only for truly astronomic values of α/β. However, if $\min(\alpha, \beta) \leqslant 1$, then $\lambda \leqslant 1$ and it is quite possible for the right-hand side of (4.27) to be detectably smaller than 1. For example, if $\beta = 0.1$, $\alpha = 0.5$, and $V = 10^{38}$, then overflow occurs if $R_1 > 0.9998$, or once every 5000 prospective variates on average. This is unacceptable. Fortunately this can be anticipated through condition (4.27), in which case the acceptance condition reduces to $4R_2R_1^2 \leqslant \{(1 + d)/d\}^{\alpha+\beta}$. If this is satisfied then $X_{\alpha, \beta} = 1$ is delivered. A further complication arises if $R_1 < U/(1 + U)$ where U is the machine's smallest positive real number. In this case G will overflow. A conservative trap for this is $R_1 < U$. This gives $Y = 0$, which is always accepted. Algorithm B1 shows the complete method.

Equation (4.25), shows that when $\min(\alpha, \beta) > 1$, $M^{-1} \geqslant e/4$ (this value occurring as $\beta \to \infty$ and $\alpha \to 1$ or vice versa), with $M \overset{\cdot}{\to} 1$ as $\alpha = \beta \to 1$. The method is therefore highly suitable for this range of parameter values. A bonus is that M remains bounded when $\min(\alpha, \beta) \leqslant 1$, the lowest sampling efficiency being 0.25 when α or $\beta \to 0$. Algorithm B1 uses simple acceptance and rejection pretests. Cheng considers more sophisticated pretests, separately for the two parameter value ranges, leading to two separate generators. In the case of $\min(\alpha, \beta) \leqslant 1$, these are so effective that the reported performance when α or $\beta \to 0$ is comparable with that when $\min(\alpha, \beta) > 1$. Algorithm B1 covers both cases in one generator, at the expense of slower performance when $\min(\alpha, \beta) \leqslant 1$, but little degradation in performance when $\min(\alpha, \beta) > 1$.

Algorithm B1

Input $[\alpha, \beta]$ Saved $[a, b, J, d, f, \lambda, T, c]$.
Output $[X]$ Parameter $[U, V =$ machine's smallest and largest positive real numbers].

S1   If $\beta > \alpha$ then
S2 $a := \beta, b := \alpha, J := 1$
S3   Else
S4 $a := \alpha, b := \beta; J := 0$
S5   End if.
S6 $d := a/b, f := a + b.$
S7   If $b > 1$ then
S8 $\lambda := \sqrt{\{(2ab - a - b)/(a + b - 2)\}}, T := 1$
S9   Else
       $\lambda := b, T := 1/\{1 + [a/(bV)]^b\}.$
S10 End If.
S11 $c := a + \lambda.$
1   generate $R_1, R_2 \sim U(0, 1)$.
2     $S := R_2 R_1^2.$
3   If $R_1 < U$ or $S \leqslant 0$ goto 1.
4   If $R_1 < T$ then
        $G := \ln\{R_1/(1 - R_1)\}/\lambda$
        $Y := d \exp(G)$
5      $Z := cG + f \ln\{(1 + d)/(1 + Y)\} - \ln 4$
6   If $S - 1 > Z$ then
7   If $S - 1 > SZ$ goto 1
8   If $\ln S > Z$ goto 1.
9   End If.
10    $X := Y/(1 + Y).$
11   Else
12   If $4S > (1 + 1/d)^f$ goto 1.
13     $X := 1$
14   End If.
15   If $J = 1$ then
16     $X := 1 - X$
17   End If.
18   Exit.

4.4.4   A family of switching algorithms

A method which is robust for all values of α and β has been proposed by Atkinson (1979a). It uses an envelope 'switching' method with target distribution

$$g_Y(x) = \begin{cases} rpx^{p-1}/t^p & (0 < x \leqslant t) \\ (1 - r)q(1 - x)^{q-1}/(1 - t)^q & (t < x \leqslant 1), \end{cases} \tag{4.28}$$

where $0 < p \leqslant \alpha$, $0 < q \leqslant \beta$, $0 \leqslant r \leqslant 1$, $0 \leqslant t \leqslant 1$. Prospective variates are generated according to

$$Y = \begin{cases} t(R_1/r)^{1/p} & (R_1 \leqslant r) \\ 1-(1-t)\{(1-R_1)/(1-r)\}^{1/q} & (R_1 > r). \end{cases}$$

From eqn (4.28)

$$h(x) = \frac{f_X(x)}{g_Y(x)} = \begin{cases} x^{\alpha-p}(1-x)^{\beta-1}t^p/\{rpB(\alpha,\beta)\} & (x \leqslant t) \\ (1-x)^{\beta-q}x^{\alpha-1}(1-t)^q/\{(1-r)qB(\alpha,\beta)\} & (x > t). \end{cases}$$

We will assume that $\alpha \leqslant \beta$ with delivery of $1 - X_{\beta,\alpha}$ if $\alpha > \beta$.

(i) $\beta \leqslant 1$

In this case $p = \alpha$, $q = \beta$. The suprema of $h(x)$ in $[0, t]$ and $(t, 1]$ occur at t and $t + \varepsilon$ (where $\varepsilon \to 0+$), respectively, leading to a choice of switching probability,

$$r = \beta t/\{\beta t + \alpha(1-t)\}$$

and

$$M = \{(1-t)\alpha + \beta t\}/\{B(\alpha+1, \beta+1)t^{1-\alpha}(1-t)^{1-\beta}\}.$$

Efficiency is maximized by minimizing M, leading to

$$t = \left[1 + \left\{\frac{\beta(1-\beta)}{\alpha(1-\alpha)}\right\}^{\frac{1}{2}}\right]^{-1} \qquad (\alpha \neq 1)$$

or $t = \frac{1}{2}$ when $\alpha = \beta = 1$.

(ii) $\cdot$ $\alpha \leqslant 1$, $\beta > 1$

As before, $p = \alpha$, $q = \beta$, but the suprema now lie at 0 and $t + \varepsilon$ leading to

$$r = Bt/\{\beta t + \alpha(1-t)^\beta\}$$

and

$$M = \frac{Bt^\alpha + \alpha(1-t)^\beta t^{\alpha-1}}{B(\alpha+1, \beta+1)}.$$

M cannot be minimized analytically but the choice

$$t = (1-\alpha)/(\beta+1-\alpha)$$

still gives a uniformly high sampling efficiency (e.g. it never drops below 0.79 when $0 < \alpha \leqslant 1$ and $1 < \beta \leqslant 10$).

(iii) $\alpha > 1$, $\beta > 1$

A natural choice for t is the mode of the distribution giving

$$t = (\alpha-1)/(\alpha+\beta-2)$$

Suprema of $h(x)$ now occur at $u = (\alpha-p)/(\alpha+\beta-p-1)$ and $v = (\alpha-1)/$

$(\alpha + \beta - q - 1)$, leading to a switching probability given by

$$\frac{1}{r} = 1 + \frac{p(1-t)^q v^{\alpha-1}(1-v)^{\beta-q}}{qt^p u^{\alpha-p}(1-u)^{\beta-1}}.$$

Optimized values for p and q require numerical methods. However, only a small loss in efficiency ($\lesssim 3\%$) occurs when using the non-optimized values,

$$p = \sqrt{(2\alpha - 1)}, \quad q = \lambda\sqrt{(2\beta - 1)} + (1 - \lambda)\beta, \quad \lambda = (2t)^{\frac{1}{2}}.$$

This does not of course affect the exactness of the procedure.
    Algorithm B2 shows an implementation of the method.

Algorithm B2

Input $[\alpha, \beta]$ Saved $[a, b, J, p, q, t, r, C_1, C_2]$
Output $[X]$

S1   If $\beta < \alpha$ then
S2 $a := \beta, b := \alpha, J := 1.$
S3   Else
S4 $a := \alpha, b := \beta, J := 0.$
S5   End If.
S6   If $b \leqslant 1$ then
S7 $p := a, q := b$
S8   If $a = 1$ then
S9 $t := \frac{1}{2}$
S10 $r := \frac{1}{2}$
S11   Else
S12 $t := 1/[1 + \sqrt{\{b(1-b)/[a(1-a)]\}}]$
S13 $r := bt/\{bt + a(1-t)\}$
S14   End If.
S15 Else if $a \leqslant 1$ then
S16 $p := a, q := b$
S17 $t := (1-a)/(b+1-a)$
S18 $r := bt/\{bt + a(1-t)^b\}$
S19 Else
S20 $t := (a-1)/(a+b-2)$
S21 $\lambda := (2t)^{\frac{1}{2}}$
S22 $p := \sqrt{(2a-1)}$
S23 $q := \lambda\sqrt{(2b-1)} + (1-\lambda)b$
S24 $u := (a-p)/(a+b-p-1)$
S25 $v := (a-1)/(a+b-q-1)$
S26 $C_1 := (a-p)\ln u + (b-1)\ln(1-u)$
S27 $C_2 := (a-1)\ln v + (b-q)\ln(1-v)$
S28 $r := \left[1 + \frac{p(1-t)^q \exp(C_2 - C_1)}{qt^p}\right]^{-1}.$
S29 End If

Algorithm B2 (Continued)

```
1   generate R_1, R_2 ~ U(0, 1)
2   If R_1 ⩽ r then
3      X := t(R_1/r)^{1/p}
4      If b⩽1 then
5         If ln R_2⩾(b−1) ln{(1−X)/(1−t)} goto 1.
6      Else if a⩽1 then
7         If ln R_2⩾(b−1) ln(1−X) goto 1
8      Else
9         If ln R_2⩾(a−p) ln X+(b−1) ln(1−X)−C_1 goto 1.
10     End If.
11  Else
12     X := 1−(1−t) {(1−R_1)/(1−r)}^{1/q}
13     If a⩽1 then
14        If ln R_2⩾(a−1) ln(X/t) goto 1
15     Else
16        If ln R_2⩾(a−1) ln X+(b−q) ln (1−X)−C_2 goto 1
17     End If.
18  End If.
19  If J=1 then
20     X := 1−X
21  End If.
22  Exit.
```

4.4.5 Comparison of methods

Table 4.4 shows marginal generation times for Jöhnk's, the log–logistic, and switching methods. Timings for the first method are shown only when $\alpha \leqslant 1$ and $\beta \leqslant 1$, as the theoretical analysis demonstrates that execution time becomes unbounded when either parameter becomes large.

No algorithm is uniformly fastest, but the switching method gives the best overall performance over the entire parameter range. Offset against this is the program length (approximately 70 FORTRAN statements) and the set-up time for 'bridge' distributions. This is of the order of 800 μs, but only 80 and 180 μs for 'U'-shaped and 'J'-shaped distributions. Jöhnk's method gives respectable performance for all 'U'-shaped distributions, and is the fastest when both parameters are small. It has the merit of extreme simplicity and no set-up calculations. The log–logistic method offers good performance except when one parameter is small. This effect will be accentuated with slower random number generators. However, it is easily implemented and unlike the switching method requires little set-up time (of the order of 80 μs). For these reasons the overall recommendation when efficiency, robustness, ease of implementation, and set-up time are all of some importance is to implement

Table 4.4 Mean time (μs) to generate one beta variate on a DEC-20 (FORTRAN implementations) when shape parameters are fixed between successive calls

$\alpha \backslash \beta$	0.1	0.5	0.9	1.0	2.0	10	100
	292‡	313	313	318	—	—	—
0.1	524	729	754	799	812	816	823
	502	382	304	416	285	290	288
		362	405	424	—	—	—
0.5		379	446	434	468	517	528
		358	317	284	326	344	351
			500	556	—	—	—
0.9			287	297	356	406	411
			302	296	336	330	331
				569	—	—	—
1.0				274	331	384	397
				175	297	280	288
					—	—	—
2.0					290	323	332
					369	381	403
						—	—
10						298	308
						401	479
							—
100							302
							429

‡ The three values given are those obtained using Jöhnk's, the log–logistic, and the switching methods, respectively.

the log–logistic method. A FORTRAN 77 function BETA is shown in Appendix 1.6. When one parameter is small and the log–logistic method degrades, a suitable back-up is the gamma method. Reference to Tables 4.1 and 4.3 show that it is faster for such values. Alternatively the more sophisticated versions of the log–logistic method suggested in Cheng's paper (1978) may be implemented.

Several other beta methods are available, some of which may be of interest for specialized values of α and β. When α and β are integer-valued, $X_{\alpha, \beta}$ may be delivered as the α order statistic in a random sample of $(\alpha + \beta - 1)$ uniform (0, 1) variates. This requires either a partial sort or recursive generation of the order statistics, given their conditional distributions (Newby, 1979). Either approach may be useful when one parameter is small and the other not too large.

'Bridge' distributions can be handled via envelope rejection with a uniform target distribution. But the sampling efficiency becomes small unless α and β

are close to 1. Ahrens and Dieter (1974) used a normal target distribution (BN). This provides a very poor fit when $\alpha = \beta \to 1$. Schmeiser and Shalaby (1980) modify BN using the squeeze technique, but the method shows little improved performance. They consider other envelope methods including a composite target distribution of six uniform and two triangular distributions. These have increased memory and set-up requirements, as does Forsythe's method. Atkinson and Pearce (1976) implemented the latter, finding set-up time to be equivalent to the generation of about 1000 variates. A further disadvantage is that the distribution has to be truncated for 'U'- or 'J'-shaped distributions.

4.5 Chi-squared, t, and F-distributions

4.5.1 Chi-squared distributions

A random variable having density

$$f_{\chi_n^2}(x) = \frac{e^{-x/2} x^{(n/2)-1}}{2^{n/2} \Gamma(n/2)} \qquad (x \geq 0),$$

where $n = 1, 2, \ldots$, has a chi-squared distribution with n degrees of freedom. Equivalently

$$\chi_n^2 = 2X_{n/2} \tag{4.29}$$

where X_α is a standard gamma variate with shape parameter α. The current availability of efficient gamma routines for $\alpha \geq \frac{1}{2}$, allows χ_n^2 to be conveniently obtained, when n is fixed or reset between calls.

Other methods based on structural properties such as

$$\chi_n^2 = \sum_{i=1}^{n} Z_i^2,$$

where $\{Z_i\}$ are independently $N(0, 1)$ are unlikely to be as good, except for small n, due to an increasing requirement for normal deviates, as n increases. A method which was frequently used for odd values of n when gamma generation was thought to be difficult for non-integral shape parameters, is to set

$$\chi_n^2 = Z^2 + 2X_{(n-1)/2} \qquad (n \geq 3).$$

Nowadays this is likely to be slower than (4.29), because of the extra normal deviate.

Non-central chi-squared variates, $S_{n,\mu}$ with non-centrality parameter μ (> 0) and n degrees of freedom may be generated through

$$S_{n,\mu} = \chi_{n-1}^2 + (Z_n - \mu^{\frac{1}{2}})^2. \tag{4.30}$$

A less obvious approach is to utilize a probability mixture representation (Johnson and Kotz, 1970b, p. 132)

$$f_{S_{n,\mu}}(x) = \sum_{j=0}^{\infty} \frac{(\frac{1}{2}\mu)^j e^{-\mu/2}}{j!} f_{\chi^2_{n+2j}}(x), \qquad (4.31)$$

returning χ^2_{n+2j} with probability that a Poisson variate (mean $\mu/2$) equals j. Such a method would be preferred to (4.30) only when Poisson generation is faster than normal generation (e.g. when μ is small). Note the requirement for a gamma generator with shape parameter reset between calls.

4.5.2 t-distribution

Variates having p.d.f.

$$f_{T_n}(x) = \frac{\left(1 + \dfrac{x^2}{n}\right)^{-(n+1)/2}}{(\sqrt{n})B\left(\dfrac{1}{2}, \dfrac{n}{2}\right)} \qquad (n = 1, 2, \ldots)$$

follow a t-distribution with n degrees of freedom. If $n = 1$ this reduces to a Cauchy distribution and inversion yields $T_1 = \tan\{\pi(R - 0.5)\}$. When $n = 2$, inversion yields $T_2 = (\sqrt{2})(R - 0.5)/\sqrt{(R - R^2)}$, as noted in Section 4.3.4. To obtain higher-order variates, we must look to more general methods. One follows directly from $T_n = Z/\sqrt{(\chi^2_n/n)}$. Although convenient this is apt to be rather slow, requiring one normal and one gamma variate. Alternatively T_n can be delivered as the square root of an F variate (see Section 4.5.3) with 1 and n degrees of freedom, ensuring that a random sign is applied. Kinderman et al. (1977) decompose the t-density over the range $(-2, 2)$ into a triangular density and a small residual density. The resulting algorithm, using envelope rejection is rather longer and not so fast as the next method to be discussed.

An easily implemented and efficient procedure has been derived by Kinderman and Monahan (1980). Using the ratio of uniforms approach (Section 3.6) the dimensions of an enclosing rectangle are

$$u_+ = \left\{(\sqrt{n})B\left(\frac{1}{2}, \frac{n}{2}\right)\right\}^{-\frac{1}{2}}$$

$$v_+ = v_- = \left\{\left(\frac{n+1}{n-1}\right)^{-(n+1)/2}\left(\frac{2n}{n-1}\right)\right\}^{\frac{1}{2}} u_+. \qquad (4.32)$$

Thus a prospective variate,

$$X = \left(\frac{2R_2 - 1}{R_1}\right)\frac{v_+}{u_+}$$

$$= \left(\frac{2R_2 - 1}{R_1}\right)g$$

where

$$g = \left(\frac{n-1}{n+1}\right)^{(n+1)/4} \left(\frac{2n}{n-1}\right)^{\frac{1}{4}}.$$

The acceptance condition becomes

$$R_1 \leqslant \left(1 + \frac{X^2}{n}\right)^{-(n+1)/4}.$$

Rewriting this in the form

$$X^2 \leqslant n\{R_1^{-4/(n+1)} - 1\},$$

acceptance and rejection pretests may be derived using bounds

$$b - aR_1 \leqslant n\{R_1^{-4/(n+1)} - 1\} \leqslant (f/R_1) - d.$$

The probability of satisfying the acceptance pretest is maximized when $b=5$ and $a = 4(1 + n^{-1})^{(n+1)/4}$. The upper bound is valid only for $n \geqslant 3$, when convenient values are $d = 3$ and $f = 16/a$. This leads to Algorithm T1.

Algorithm T1

Input $[n]$ Saved $[c, a, f, g]$
Output $[X]$

S1 $c := -(n+1)/4$
S2 $a := 4(1 + n^{-1})^{-c}$
S3 $f := 16/a.$
S4   If $n = 1$ then
S5 $g := 1$
S6   Else

S7 $g := \left(\frac{n+1}{n-1}\right)^c \left(\frac{2n}{n-1}\right)^{\frac{1}{4}}.$

S8   End If.

1   generate $R_1, R_2 \sim U(0, 1)$.
2    $X := (2R_2 - 1)g/R_1$.
3   If $X^2 \leqslant 5 - aR_1$ exit
4   If $n \geqslant 3$ and $R_1(X^2 + 3) > f$ goto 1
5   If $R_1 > (1 + X^2/n)^c$ goto 1
6   Exit.

Results (4.32) show that the sampling efficiency

$$P = (4u_+ v_+)^{-1}$$

$$= \frac{(\sqrt{n})B\left(\frac{1}{2}, \frac{n}{2}\right)}{4} \left(\frac{n+1}{n-1}\right)^{(n+1)/4} \left(\frac{n-1}{2n}\right)^{\frac{1}{4}},$$

which ranges from $\pi/4$ at $n=1$ to an asymptotic value (obtainable from the asymptotic normality) of $\sqrt{(\pi e)}/4 = 0.731$ as $n \to \infty$. Using a FORTRAN implementation on a DEC-20, under the conditions stated in Section 4.3.5, marginal generation time was found to be 157, 134, 103, 112, 112, 112, 110, and 113 μs at $n=1, 2, 3, 5, 10, 20, 50$, and 100, respectively. When n is reset between calls these times increase by approximately 300 μs (excepting $n=1$, where the increase is 140 μs). The high acceptance probability, together with effective pretests, results in a generator which is not very much slower than the best portable normal generators. For this reason it is recommended in preference to other methods, and a FORTRAN 77 function TDISTN appears in Appendix 1.7.

A non-central t random variable with n degrees of freedom and non-centrality parameter δ is defined by

$$T_{n,\delta} = \frac{Z+\delta}{\sqrt{(\chi_n^2/n)}}$$

where Z and χ_n^2 are independent standard normal and chi-squared with n degrees of freedom random variables. This provides a convenient method for generating such variates. A doubly non-central t random variable with non-centrality parameters δ and μ is defined by

$$T_{n,\delta,\mu} = \frac{Z+\delta}{\sqrt{(S_{n,\mu}/n)}}$$

where $S_{n,\mu}$ is a non-central chi-squared random variable with n degrees of freedom and non-centrality parameter μ. This relationship may be used to generate such variates. Alternatively, since $S_{n,\mu}$ is a Poisson mixture of central chi-squared random variables (eqn 4.31), we may deliver

$$T_{n,\delta,\mu} = \left(\sqrt{\frac{n}{n+2j}}\right) T_{n+2j,\delta}$$

with probability that a Poisson random variable (mean $\mu/2$) equals j.

4.5.3 F-distribution

A random variable F_{n_1,n_2} having p.d.f.

$$f_{F_{n_1,n_2}}(x) = \frac{n_1^{n_1/2} n_2^{n_2/2} x^{(n_1/2)-1}}{B(n_1/2, n_2/2)(n_2+n_1 x)^{(n_1+n_2)/2}},$$

where $x \geqslant 0$ and $n_1, n_2 = 1, 2, \ldots$, follows an F-distribution with n_1 and n_2 degrees of freedom. Equivalently

$$F_{n_1,n_2} = \frac{n_2 X}{n_1(1-X)}$$

where X is a beta variate with parameters $\alpha = n_1/2$ and $\beta = n_2/2$. Thus, just as the chi-squared distribution requires an efficient gamma generator, so the F-distribution requires a good beta procedure. The log–logistic method (Section 4.4.3) is recommended for this task. Equation (4.25) indicates that the sampling efficiency is at least $e/4 = 0.680$ when n_1 and $n_2 \geqslant 2$. When one parameter (n_1 say) is 1, the efficiency is $\pi/4$ at $n_2 = 1$, $(\sqrt{3}/2)^3 = 0.650$ at $n_2 = 2$, approaching an asymptotic value of $\sqrt{\{(\pi e)/32\}} = 0.517$ as $n_2 \to \infty$. Alternatively when $n_1 = 1$, a good method is to return the square of a t-variate, $T^2_{n_2}$, obtained from Algorithm T1. A non-central F random variable with n_1 and n_2 degrees of freedom and non-centrality parameters μ_1 and μ_2 is defined by

$$F_{n_1, n_2} = \frac{\sqrt{(S_{n_1, \mu_1}/n_1)}}{\sqrt{(S_{n_2, \mu_2}/n_2)}}$$

where S_{n_1, μ_1} and S_{n_2, μ_2} are independent, non-central, chi-squared random variables. Using either (4.30) or (4.31) this provides a method for generating such variates.

4.6   Some other continuous distributions

4.6.1   Easily derived procedures

A random variable has a log–normal distribution if its p.d.f. is of the form

$$f_X(x) = \frac{e^{-(\ln x - \mu)^2/2\sigma^2}}{\sqrt{(2\pi)}\sigma x} \qquad (x \geqslant 0),$$

where μ and $\sigma (>0)$ are constants. Structurally, $X = e^Y$ where $Y \sim N(\mu, \sigma^2)$, and this provides an obvious method of generation.

The standard Cauchy distribution has p.d.f.,

$$f_X(x) = \{\pi(1 + x^2)\}^{-1}.$$

Generation is possible using inversion, the ratio of two standard normal deviates, or delivering a t-variate with one degree of freedom, the latter being recommended if Algorithm T1 is used.

Generation from a three-parameter Weibull distribution is conveniently managed using inversion, as described in Section 3.1.

A Laplace or double exponential distribution has p.d.f.

$$f_X(x) = \tfrac{1}{2}\lambda e^{-\lambda|x - \beta|} \qquad (\lambda > 0)$$

and inversion provides a direct generation method.

The logistic distribution has density

$$f_X(x) = \frac{e^{-(x - \alpha)/\beta}}{\beta\{1 + e^{-(x - \alpha)/\beta}\}^2} \qquad (\beta > 0),$$

and inversion yields

$$X = \alpha + \beta \ln \{R/(1-R)\}.$$

The Pareto distribution has p.d.f.

$$f_X(x) = \frac{ak^a}{x^{a+1}} \qquad (x \geq k),$$

where $a > 0$ and $k > 0$. Inversion gives $X = kR^{-1/a}$.

4.6.2 The generalized inverse Gaussian distribution

A distribution which, at the time of writing, is not widely known is the generalized inverse Gaussian (Jørgensen, 1982). This has density

$$f_{X'}(x) = \frac{(\psi/\chi)^{\lambda/2} x^{\lambda-1} e^{-\frac{1}{2}(\chi x^{-1} + \psi x)}}{2K_\lambda(\sqrt{(\chi\psi)})} \qquad (x \geq 0), \qquad (4.33)$$

where $K_\lambda(\cdot)$ is the modified Bessel function of the third kind with index λ. The domain of variation of the parameters is $\chi > 0$, $\psi \geq 0$ if $\lambda < 0$; $\chi > 0$, $\psi > 0$ if $\lambda = 0$; $\chi \geq 0$, $\psi > 0$ if $\lambda > 0$. Atkinson (1982) describes the density as an enriched family of gamma distributions. For this reason it is likely that the distribution will be used frequently in future. Additionally it has been mixed with a normal distribution to provide the generalized hyperbolic distribution. A special case of the latter occurs when $\lambda = 1$, the hyperbolic distribution. This has been used (Barndorff–Nielsen, 1977, 1978, 1979) to model the variation in size of wind-blown sand particles, diamond sizes, and incomes. Special cases of the generalized inverse Gaussian are the gamma distribution ($\chi = 0$), the reciprocal gamma distribution ($\psi = 0$), the inverse Gaussian distribution ($\lambda = -\frac{1}{2}$), and the reciprocal inverse Gaussian or random walk distribution ($\lambda = \frac{1}{2}$).

Atkinson devised a composition/envelope rejection scheme, in which the target distribution consists of a mixture of three non-overlapping densities. Parameter determination for this target distribution requires numerical procedures. The result is a substantial algorithm, with low efficiency for certain parameter values, which the experimenter may find difficult to implement.

Fortunately a simple alternative method can be derived, using the ratio method. We will exclude the cases $\chi = 0$, $\psi = 0$, since efficient gamma (and hence reciprocal gamma) generators already exist. Reparameterizing (4.33), set $\alpha = \sqrt{(\psi/\chi)}$ and $\beta = \sqrt{(\psi\chi)}$, and define $X = X'\alpha$. Then the distribution reduces to a two-parameter family,

$$f_{X,\lambda}(x) = \frac{x^{\lambda-1} e^{-\beta(x+x^{-1})/2}}{2K_\lambda(\beta)} \qquad (x \geq 0).$$

It is only necessary to consider $\lambda \geqslant 0$, since the reciprocal variate $Y = X^{-1}$ belongs to the same family with λ replaced by $-\lambda$. The mode of X lies at

$$m_X(\lambda) = \frac{\lambda - 1 + \sqrt{\{(\lambda - 1)^2 + \beta^2\}}}{\beta},$$

and that of Y at $m_Y(\lambda) = m_X(-\lambda)$. Thus the dimensions of an enclosing rectangle for the acceptance region are

$$u_+ = f_{X, \lambda}^{\frac{1}{2}}(m_X(\lambda))$$

and

$$v_+ = f_{X, -\lambda}^{\frac{1}{2}}(m_X(-\lambda)),$$

leading to a prospective variate

$$X = \frac{v_+}{u_+} \frac{R_2}{R_1}$$

and an acceptance condition,

$$R_1 \leqslant \left[\frac{X}{m_X(\lambda)}\right]^{(\lambda - 1)/2} \exp\{-\beta(X + X^{-1} - m_X(\lambda) - [1/m_X(\lambda)])/4\}.$$

The resulting method is shown in Algorithm GIG1. Note that for large λ/β, $m_X/m_Y \to (2\lambda/\beta)^2$ which may overflow. Statement S1 is a conservative trap for this condition. The acceptance probability $P = (2u_+ v_+)^{-1}$ or

$$P = K_\lambda(\beta)[m_X(\lambda)]^{(1 - \lambda)/2}[m_X(-\lambda)]^{(1 + \lambda)/2}$$

$$\times \exp(0.25\beta\{m_X(\lambda) + [1/m_X(\lambda)] + m_X(-\lambda) + [1/m_X(-\lambda)]\}).$$

Specimen values of P are shown in Table 4.5. The efficiency is not uniformly

Algorithm GIG1 $(\lambda \geqslant 0, \beta > 0)$

Input $[\lambda, \beta]$ Saved $[A, C]$
Output $[X]$ Parameter $[V = \text{machine's largest real number}]$

S1   If $\lambda/\beta > 0.25\sqrt{V}$ stop.
S2 $m_X := [\lambda - 1 + \sqrt{\{(\lambda - 1)^2 + \beta^2\}}]/\beta$.
S3 $m_Y := [-\lambda - 1 + \sqrt{\{(\lambda + 1)^2 + \beta^2\}}]/\beta$.

S4 $A := (m_X m_Y)^{-\lambda/2}\left(\dfrac{m_X}{m_Y}\right)^{\frac{1}{4}} \exp[0.25\beta(m_X + m_X^{-1} - m_Y - m_Y^{-1})]$.

S5 $C := -0.5(\lambda - 1)\ln m_X + 0.25\beta(m_X + m_X^{-1})$.
1   generate $R_1, R_2 \sim U(0, 1)$.
2     $X := AR_2/R_1$.
3   If $\ln R_1 > 0.5(\lambda - 1)\ln X - 0.25\beta(X + X^{-1}) + C$ goto 1.
4   Exit.

Table 4.5 Sample efficiencies for generalized inverse Gaussian generation (ratio method)

$\lambda \backslash \beta$	0.2	0.4	1	5	10	20
0	0.481	0.630	0.717	0.496	0.373	0.272
0.5	0.555	0.667	0.719	0.495	0.372	0.272
1	0.791	0.733	0.717	0.493	0.372	0.272
2	0.709	0.703	0.674	0.484	0.370	0.271
5	0.515	0.515	0.510	0.438	0.356	0.268
10	0.381	0.380	0.380	0.360	0.321	0.258
20	0.275	0.274	0.274	0.270	0.260	0.231

high, which is the price paid for the ease with which the algorithm is implemented. It should be adequate, however, for most experimental purposes. A FORTRAN 77 function GIG appears in Appendix 1.8. On a DEC-20, marginal generation time was 207 μs when $\lambda = 0.5$ and $\beta = 1$, with timings for other parameter values correlating fairly closely with the sample efficiency. The set-up time was of the order of 380 μs, regardless of parameter values. For the special cases $\lambda = \frac{1}{2}$, $-\frac{1}{2}$, it is probably better to use Algorithm 3.9, discussed previously. This gave marginal generation times of the order of 160 μs on a DEC-20, regardless of the β-value.

4.6.3 von Mises distribution

A random variable has the von Mises distribution if its p.d.f. is

$$g_{\Theta}(\theta) = \{2\pi I_0(k)\}^{-1} e^{k \cos(\theta - \mu_0)}$$

where $-\pi \leqslant \theta < \pi$, $-\pi \leqslant \mu_0 < \pi$, $k > 0$, and $I_0(k)$ is a modified Bessel function of the first kind and order zero. Without loss of generality we may take $\mu_0 = 0$, and consider a folded distribution with p.d.f.

$$f_{\Theta}(\theta) = \{\pi I_0(k)\}^{-1} e^{k \cos \theta} \qquad (0 \leqslant \Theta \leqslant \pi).$$

Since Θ is an angular random variable, the distribution has proved useful in modelling directional data, examples appearing in Mardia (1972, 1975) and Kendall (1974).

Best and Fisher (1979) proposed variate generation using envelope rejection with a wrapped Cauchy target distribution. They found this to be faster than a uniform target distribution method, suggested by Seigerstetter (1974). Dagpunar (1983) developed an algorithm using Forsythe's method. For $k \leqslant 0.5$, only one interval is required and the procedure was the fastest of all methods investigated, whether k was fixed or reset between calls. When $k > 0.5$, several intervals are required, necessitating numerical integrations.

Although faster when k is fixed between calls the numerical component of this algorithm makes it more difficult to implement. Thus Best and Fisher's method is recommended when reasonable speed and ease of implementation are both of importance.

Best and Fisher's method uses a folded wrapped Cauchy distribution,

$$g_Y(\theta) = (1-\rho^2)/\{\pi(1+\rho^2-2\rho\cos\theta)\} \qquad (0 \leqslant \theta \leqslant \pi),$$

where ρ is free to be chosen in $[0, 1)$. The distribution function for Y is

$$G_Y(\theta) = \pi^{-1}\cos^{-1}\left\{\frac{(1+\rho^2)\cos\theta - 2\rho}{1+\rho^2 - 2\rho\cos\theta}\right\} \qquad (0 \leqslant \theta \leqslant \pi).$$

Using inversion, $G_Y(Y) = R_1$. Prospective variates are therefore obtained via

$$\cos Y = (1+rz)/(r+z)$$

where $r = (1+\rho^2)/(2\rho)$ and $z = \cos\pi R_1$. The optimal value of ρ, and the associated sampling efficiency, M^{-1}, are determined by

$$M = \min_{\rho} \max_{\theta} \{f_\Theta(\theta)/g_Y(\theta)\}.$$

The corresponding (θ, ρ) is therefore a saddle point of $f_\Theta(\theta)/g_Y(\theta)$, and is given by

$$\cos\theta_0 = \{1+\rho^2 - (2\rho/k)\}/2\rho$$

and the (single) root of

$$-k\rho^4 + 2\rho^3 + 2k\rho^2 + 2\rho - k = 0,$$

which lies in $[0, 1)$. Thus $\rho = \{\tau - \sqrt{(2\tau)}\}/2k$ where $\tau = \sqrt{(4k^2+1)} + 1$. The acceptance condition is

$$R_2 < e^{k(\cos Y - \cos\theta_0)}\left(\frac{1+\rho^2 - 2\rho\cos Y}{1+\rho^2 - 2\rho\cos\theta_0}\right)$$
$$= ce^{-c+1},$$

where $c = k(r - \cos Y)$. A suitable acceptance pretest is $R_2 < c(2-c)$. For optimized ρ, the acceptance probability, M^{-1}, approaches 1 as $k \to 0$, and $\sqrt{\{e/(2\pi)\}} = 0.658$ as $k \to \infty$, indicating that the method is robust for all parameter values. Marginal generation time on a DEC 20 ranges from approximately 200 μs at $k = 0$, to 280 μs at $k = 20$. Set-up time is approximately 80 μs. A small improvement can often be obtained by using a 'polar' method for computing $\cos\theta$. Algorithm VM giving a version in which $\cos\theta$ is evaluated conventionally, and a random sign is applied to the folded variate. Step 5 is an acceptance pre-test. A FORTRAN 77 function VMISES appears in Appendix 1.9. When implemented on a DEC-20 computer this was at most 5% slower than a version using the polar modification.

Algorithm VM

Input [k]   Saved [r]
Output [X]

S1 $\tau := 1 + (1 + 4k^2)^{\frac{1}{2}}$.
S2 $\rho := \{\tau - \sqrt{(2\tau)}\}/(2k)$.
S3 $r := (1 + \rho^2)/(2\rho)$.
1   generate R_1, $R_2 \sim U(0, 1)$.
2      $z := \cos \pi R_1$.
3      $f := (1 + rz)/(r + z)$.
4      $c := k(r - f)$.
5   If $R_2 \geqslant c(2 - c)$ then
6   If $\ln(c/R_2) + 1 - c < 0$ goto 1.
7   End If.
8      $X := \cos^{-1} f$.
9   generate $R_3 \sim U(0, 1)$.
10   If $R_3 < 0.5$ then
11        $X := -X$.
12   End If.
13   Exit.

5
Discrete distributions

The generation of discrete variates poses difficulties not found in the continuous case. The treatment of factorials or binomial coefficients requires particular care to avoid overflow. While envelope rejection can be used with a discrete target distribution, the latter is limited to uniform and geometric distributions if ease of implementation is required, since only these can be inverted analytically. Lastly, the ratio of uniforms method, which has proved so useful for continuous variates, cannot be applied in any obvious sense to discrete distributions.

5.1 Poisson distribution

Methods of Poisson variate generation are important because of the predominance of the Poisson process in natural phenomena. The p.m.f. is

$$f_X(x) = \frac{\mu^x e^{-\mu}}{x!} \qquad (x \geq 0)$$

where $\mu > 0$. Comparative timings for many of the methods to be discussed are given in Atkinson (1979b, c).

Probably the most familiar method is the multiplicative approach, described in Section 3.3. This becomes inefficient for large μ. For small values of $\mu (\lesssim 5)$, particularly with a fast uniform generator, it can be competitive, being reasonably fast, easy to implement, and flexible with changing μ.

An alternative is inversion using either stored or unstored versions (Algorithms 3.10 and 3.12, respectively). The latter is more suitable when μ is reset between calls. Although the number of comparisons varies as μ, the method can be competitive for moderate values ($\mu \lesssim 15$), because of its ease of implementation. Algorithm P1 gives an unstored version, with a trap for the underflow of $e^{-\mu}$.

At the cost of some precalculation, the number of comparisons can be made to vary as $O(\mu^{\frac{1}{2}})$ by starting at the mode, as in Fishman (1976b) and Kemp (1982). Note that in these cases, care is needed in evaluating the modal probability $f_X(\langle \mu \rangle)$, when μ is large. Expressing this in the form

$$f_X(\langle \mu \rangle) = (\mu/e)^{\langle \mu \rangle} e^{-\mu + \langle \mu \rangle} / \langle \mu \rangle!,$$

Algorithm P1

Input [μ]   Saved [W]
Output [X]   Parameter [U = machine's smallest positive real number].

S1   If $\mu > -\ln U$ stop.
S2 $W := \exp(-\mu)$.
1   generate $R \sim U(0, 1)$, $X := 0$, $r := W$.
2   While $r < R$ do
3          $R := R - r$
4          $X := X + 1$
5          $r := r\mu/X$
6   End While.
7   Exit.

and taking a machine with a hypothetical floating point overflow at 10^{38}, $(\mu/e)^{\langle \mu \rangle}$ and $\langle \mu \rangle!$ will overflow at $\mu = 35$ and 34, respectively. The remedy is to evaluate

$$\ln[f_X(\langle \mu \rangle)] = \langle \mu \rangle \ln \mu - \mu - \ln \langle \mu \rangle!$$

$$\simeq \langle \mu \rangle \ln(\mu/\langle \mu \rangle) + \langle \mu \rangle - \mu - \tfrac{1}{2}\ln\langle \mu \rangle - \{12\langle \mu \rangle\}^{-1} - \ln \sqrt{(2\pi)},$$

and then exponentiate the result. This is based on Stirling's approximation and gives good accuracy (better than one part in 10^6 for $\mu > 15$). Using a stored version of modal search, Atkinson found that even at $\mu = 70$ the method was no less than half the speed of the fastest algorithm available (the alias method).

Ahrens and Dieter (1980) have used envelope rejection with a Laplace target distribution. The resulting exact method has a bounded execution time. They found inversion to be faster when $\mu \leqslant 12$. Another method with similar characteristics is to sample a normal variate with high probability and a variate from the residual density with low probability, taking the integer part of such variates. This method (PD) given by Ahrens and Dieter (1982) is reputedly the fastest of all known methods, but requires extensive coding. Ahrens and Dieter (1974) also describe the 'Centre Triangle' algorithm (PT). If μ is sufficiently small (<9) the multiplicative method is used. Otherwise sampling is via the addition of Poisson($\langle \mu \rangle$) and Poisson($\mu - \langle \mu \rangle$) variates. The former is generated via a composition method with triangular and residual densities. The method requires considerable setting up of constants and tables, marginal generation time behaving as $O(\mu^{\frac{1}{2}})$. In the same paper Ahrens and Dieter show that a Poisson variate may be decomposed into either a Poisson variate (having smaller mean) or a binomial variate. A gamma generator is also required and marginal generation time becomes unbounded as $\mu \to \infty$. The reader may be left with the impression that there are no methods which are easily coded and at the same time have reasonable efficiency for a wide range of μ-values. However, Atkinson (1979b) described a method using a logistic envelope. Since few lines of program are needed to implement the algorithm, the generator would be portable were it not for the requirement to store a table of log factorials. Fortunately this problem may be eliminated by incorporating pretests in the following modified version of Atkinson's method.

Poisson(μ) variates, N, may be obtained from continuous variates X, having density

$$f_X(x) = \frac{\mu^{\langle x+0.5\rangle}e^{-\mu}}{\langle x+0.5\rangle!} \qquad (x > -0.5), \tag{5.1}$$

by setting $N = \langle X + 0.5 \rangle$. Given eqn (5.1), a suitable target distribution is logistic, truncated at $x = -0.5$ and having c.d.f.

$$G_Y(x) = \frac{1 - e^{-\beta(x+0.5)}}{1 + e^{-(\beta x - \alpha)}} \qquad (x > -0.5).$$

By choosing $\beta = \pi/\sqrt{(3\mu)}$ and $\alpha = \beta\mu$, the mean and variance of the Poisson and *untruncated* logistic are made identical. Given a random number R_1, inversion yields

$$R_1 = \frac{1 - e^{-\beta(X+0.5)}}{1 + e^{-(\beta X - \alpha)}},$$

or

$$X = -0.5 + \beta^{-1}\ln\left\{\frac{1 + R_1 e^{\alpha+0.5\beta}}{1 - R_1}\right\}.$$

Given a random number R_2, the acceptance condition is

$$R_2 M g_Y(X) \leqslant \frac{\mu^{\langle X+0.5\rangle}e^{-\mu}}{\langle X+0.5\rangle!}, \tag{5.2}$$

where

$$M \geqslant \max[f_X(x)/g_Y(x)].$$

Since

$$g_Y(X) = \frac{dR_1}{dX}$$

$$= \frac{\beta(1 - R_1)(1 + R_1 e^{\alpha+0.5\beta})}{(1 + e^{\alpha+0.5\beta})},$$

(5.2) becomes

$$\frac{\beta R_2(1 - R_1)(1 + R_1 e^{\alpha+0.5\beta})M}{1 + e^{\alpha+0.5\beta}} \leqslant \frac{\mu^N e^{-\mu}}{N!}.$$

As $N!$ can easily overflow, it is necessary to take logs giving

$$\ln\{R_2(1 - R_1)(1 + R_1 e^{\alpha+0.5\beta})\} + \ln(N!/\mu^N) \leqslant -\mu + \ln\{(1 + e^{\alpha+0.5\beta})/M\beta\}. \tag{5.3}$$

A major problem with this acceptance condition is the evaluation of $S_N = \ln(N!/\mu^N)$. A solution which retains compactness and speed is to use effective pretests based on the Stirling bounds,

$$0.5\ln(2\pi) + (N+0.5)\ln N - N \leqslant \ln N! \leqslant 0.5\ln(2\pi) + (N+0.5)\ln N - N + (12N)^{-1}$$

Using these in condition (5.3), acceptance and rejection pretests $(N>0)$ become

$$\ln\{R_2(1-R_1)(1+R_1 e^{\alpha+0.5\beta})\} + 0.5\ln(2\pi\mu) + (N+0.5)\ln(N/\mu)$$
$$- N + (12N)^{-1} \leqslant -\mu + \ln\{(1+e^{\alpha+0.5\beta})/M\beta\}$$

and

$$\ln\{R_2(1-R_1)(1+R_1 e^{\alpha+0.5\beta})\} + 0.5\ln(2\pi\mu) + (N+0.5)\ln(N/\mu)$$
$$- N > -\mu + \ln\{(1+e^{\alpha+0.5\beta})/M\beta\},$$

respectively. Due to the tightness of these bounds (even for small N), there is only a very small probability that S_N will have to be evaluated (expensively) via the recursion $S_N = \ln(N/\mu) + S_{N-1}$. Algorithm P2 shows how the method can be implemented.

Algorithm P2

Input $[\mu, M]$ Saved $[\beta, \gamma, \Delta]$.
Output $[N]$

S1 $\beta := \pi/\sqrt{(3\mu)}$, $\gamma := \exp[\beta(\mu+0.5)]$.
S2 $\Delta := -\mu + \ln\{(1+\gamma)/M\} + 0.5\ln(1.5/\pi^3)$.
1   generate $R_1, R_2 \sim U(0,1)$.
2    $N := \langle \beta^{-1}\ln\{(1+\gamma R_1)/(1-R_1)\}\rangle$.
3    $W := \ln\{R_2(1-R_1)(1+\gamma R_1)\}$.
4    $V := 0$.
5   If $N>0$ then
6      $Z := W + (N+0.5)\ln(N/\mu) - N$.
7   If $Z + (12N)^{-1} \leqslant \Delta'$ exit.
8   If $Z > \Delta$ goto 1
9      $V := \ln(N!/\mu^N)$
10   End If.
11   If $W + V > \Delta + 0.5\ln(2\pi\mu)$ goto 1
12   Exit.

It is clear that the maximum value of $f_X(x)/g_Y(x)$ is obtained by setting

$$M = e^{-(\mu+T)}(1+e^{\alpha+0.5\beta})/\beta$$

where

$$T = \min_{0\leqslant R_1 <1} [\ln\{(1-R_1)(1+R_1 e^{\alpha+0.5\beta})\} + \ln(N!/\mu^N)]$$

and

$$N = \left\langle \beta^{-1} \ln\left[\frac{1 + R_1 e^{\alpha + 0.5\beta}}{1 - R_1} \right] \right\rangle.$$

Table 5.1 gives estimates of M obtained numerically. For values of μ not shown in the table, the value used is the one corresponding to the next smallest μ. This incurs a small loss in efficiency, but still produces 'exact' variates.

Table 5.1 Reciprocal of sampling efficiency for logistic Poisson method

μ	0.5	1	2	5	10	20	30	50	100	200	1000
M	17.48	4.854	2.914	2.049	1.700	1.546	1.484	1.421	1.360	1.323	1.276

Table 5.2 shows execution times for Algorithms P1 and P2, with μ fixed and reset between calls. Inversion is not used for $\mu \geq 100$, since underflow of $e^{-\mu}$ is likely on most machines. The logistic method is not used for $\mu < 1$, since the sampling efficiency is very poor. Experiments were carried out on a DEC-20 (FORTRAN implementation using RAN) and on a BBC micro-computer (BASIC implementation using RND uniform generator). The apparent anamolous rise in reset times for the BBC as μ increases from 100 to 1000 is due to the exponential evaluation (required for γ) being much slower for large argument values.

Finally, which of the many Poisson generators should the experimenter use? As always the choice depends on the criteria set. If marginal generation time is all important, regardless of set-up time and program complexity, Atkinson's experiments show that the alias, indexed search, and Norman–Cannon methods are all very fast. Section 5.4 gives an alias algorithm which avoids truncation of the distribution. This results in marginal generation times of approximately 55 μs on a DEC-20, regardless of μ, with a set-up time of 5200 μs when $\mu = 50$. For ease of implementation, low storage requirements, and robustness with respect to μ, Table 5.2 indicates that unstored inversion could be used for values of μ up to 15/20 (μ fixed) or 20/30 (μ reset), and the logistic method for higher values. A FORTRAN 77 function IPOIS appears in Appendix 1.10. This calls the function IPOIS1 (unstored inversion) when $\mu < 20$ and IPOIS2 (logistic method) when $\mu \geq 20$.

5.2 Binomial distribution

$X_{n,\,p}$ is Binomially distributed if its p.m.f. is

$$f_{X_{n,\,p}}(x) = \binom{n}{x} p^x q^{n-x} \qquad (x = 0, 1, \ldots, n),$$

Table 5.2 Mean time to generate one Poisson variate on a DEC-20 (μs) and a BBC microcomputer (ms) using inversion and logistic methods (times for μ reset are shown in parentheses)

μ	0.5	1	5	10	20	50	100	1000
DEC-20								
Inversion	38(97)	47(107)	108(172)	178(238)	321(404)	761(836)	—	—
Logistic	—	931(1156)	455(676)	387(609)	345(599)	320(567)	310(553)	294(542)
BBC								
Inversion	9(27)	14(21)	48(62)	89(110)	174(213)	439(516)	—	—
Logistic	—	351(487)	177(264)	135(234)	129(223)	114(223)	112(224)	117(278)

where $0 \leqslant p \leqslant 1$ and $q = 1 - p$. A well-known generation method is to simulate n Bernoulli trials, p being the success probability per trial. Only when n is small can such a method be considered competitive. An alternative is to use numerical inversion. Behaviour will be similar to the Poisson case. If $p > \frac{1}{2}$, it is better to return $n - X_{n,\,q}$, bringing the mean number of comparisons down to $1 + \{n \min(p, q)\}$. Algorithm B gives the details. Given a machine with floating point underflow at 10^{-US}, q'^n will underflow if $np' > -p' US \ln 10 / \ln q'$. The right-hand side of this inequality attains its lowest value of $1.66096US$ at $p' = 0.5$, which provides the trap for underflow in S6. This should be large enough for most machines, since the algorithm would be unsuitable in any case, on account of its speed, when np' is very large.

Algorithm B

```
Input [n, p]      Saved [J, W, H]
Output  [X]   Parameters  [US = −log₁₀ (machine's smallest positive real
                          number)]
S1   If p ⩽ 0.5 then
S2       p′:= p, q′:= 1 − p, J:= 0
S3   Else
S4       p′:= 1 − p, q′:= p, J:= 1.
S5   End If.
S6   If np′ > 1.66096US stop.
S7   W:= q′ⁿ, H:= p′/q′.
1      generate R ∼ U(0, 1), X:= 0, Y:= W.
2      While Y < R do
3          R:= R − Y
4          X:= X + 1
5          Y:= YH(n − X + 1)/X
6      End While.
7      If J = 1 then
8          X:= n − X
9      End If.
10     Exit.
```

Since the domain of $f_X(\cdot)$ is bounded, a simple form of envelope rejection with a discrete uniform target distribution, $g_Y(x) = 1/(n+1)$ is also possible. For large n the asymptotic normality gives

$$M = \max[f_X(x)/g_Y(x)]$$
$$\simeq \sqrt{\{n/(2\pi pq)\}}$$

indicating that marginal generation time is unbounded as $n \to \infty$.

Relles (1972) exploits a relationship between binomial and beta distributions. Let V be a beta variate with parameters j and $n-j+1$, respectively, for $j \in (1, 2, \ldots, n)$. Then

$$X_{n,p} = \begin{cases} j + X_{n-j,\,(p-V)/(1-V)} & (V \leqslant p) \\ X_{j-1,\,p/V} & (V > p) \end{cases} \tag{5.4}$$

and

$$X_{0,\cdot} = 0.$$

The recursion terminates when $X_{0,\cdot}$ is called for. In this way $X_{n,p}$ may be generated solely through a sequence of beta variates. The choice of j (at any given stage in the recursion) is left to the implementer of the algorithm. In algorithms RBINOM (Relles) and BB (Ahrens and Dieter, 1974) j is set to $(n+1)/2$ if n is odd-valued. If n is even-valued, a single Bernoulli variate is generated and added to $X_{n-1,p}$. Although an elegant method, it requires an efficient beta generator, and still has an unbounded requirement for such variates as n increases. Fishman (1979) found that it is competitive only for a restricted range of p values when $n \lesssim 5$. One seemingly unnoticed virtue of the method is that the decomposition (5.4) provides an elegant way of generating truncated binomial variates. To generate $X_{n,p}$ subject to $X_{n,p} \geqslant j$, sample the initial beta variate subject to $V \leqslant p$. Conversely, generation subject to $X_{n,p} < j$ is via the generation of an initial beta variate, subject to $V > p$.

A disadvantage of methods described above is that in each case marginal generation time becomes unbounded as $n \to \infty$. Fishman (1979) gives an envelope rejection procedure with bounded execution times. The target distribution is Poisson, mean μ, which is optimized when

$$\mu = \begin{cases} n - \langle nq \rangle & (nq - \langle nq \rangle \leqslant p) \\ p(\langle nq \rangle + 1)/q & (\text{otherwise}). \end{cases}$$

If $p > \frac{1}{2}$ then $n - X_{n,q}$ is returned. The proportion of Poisson variates which are accepted is never less than $1/\sqrt{2}$. The method (BP) requires a table of $(n+1)$ logarithms to perform the acceptance test, plus an exponential and Poisson variate. For large μ, Fishman used a normal approximation to the Poisson, though clearly a bounded Poisson method such as the logistic could now be used. Fishman found the method was competitive only when n is large ($\gtrsim 20$) and $\min(p, q)$ not close to zero. For most of the remaining cases inversion proved fastest except when n was small ($\lesssim 5$) and $\min(p, q)$ not close to zero. In these cases the beta method was fastest.

Ahrens and Dieter (1974) have developed a method (algorithm BC), based on the representation of the success probability as a sequence of random bits. Envelope rejection figures in an exact method due to Ahrens and Dieter (1980), where a Laplace target distribution is used. Tail values outside $[0, n]$ are rejected. The Laplace envelope ensures that the sampling efficiency approaches $\sqrt{\{\pi/(2e)\}}$ as $\mu \to \infty$, corresponding to the normal approximation to the binomial. The method requires a rather lengthy code and inversion proved to be faster if $n \min(p, q) < 16$, while Fishman's BP (exact version without normal approximation) was faster when $np < 100$.

In recommending a binomial generator, algorithm B is suitable if $n \min(p, q)$ is not large. A FORTRAN 77 function IBINOM appears in Appendix 1.13. For other values of n and p, Section 5.4 gives an alias implementation which gives very low marginal generation times at the expense of increased memory and set-up requirements.

5.3   Negative binomial distribution

This has p.m.f.

$$f_{X_{p, k}}(x) = \frac{\Gamma(x+k)p^x q^k}{\Gamma(k)\Gamma(x+1)} \qquad (x = 0, 1, \ldots)$$

where $0 < p < 1$, $q = 1 - p$, and $k > 0$. It is in effect the discrete analogue of the gamma distribution. For integral k, $X + k$ is the sum of k i.i.d. geometric variates, each with mean q^{-1}. Thus one generation method is to simulate Bernoulli trials till the kth failure occurs. Such a method requires a mean of k/q random numbers. If k/q is large, but k is not, then a better method is to generate k geometric variates, requiring k logarithmic evaluations.

Neither of these methods can be used directly if k is non-integral. In these cases unstored inversion can be employed, requiring a mean of $1 + kp/q$ comparisons. If p is sufficiently large then it is faster (for $k \geqslant 1$) to use the reproductive property of the distribution and to deliver $X_{p, k}$ as $X_{p, \langle k \rangle} + X_{p, k - \langle k \rangle}$, the former obtained using $\langle k \rangle$ logarithmic evaluations, and the latter using unstored inversion. Algorithm NB shows such a method, in which the demarcation value of p is H. Experiments on DEC-20 and BBC computers indicated that for a wide range of k-values, $H = 0.7$ is a satisfactory choice for both machines. In step 4 $\langle k \rangle$ random numbers are required. Table 5.3 shows timings when k and p are reset between calls. Some reduction may be expected if $\ln p$ and q^k can be saved between calls.

A relationship between gamma, Poisson, and negative binomial variates provides a method with bounded execution times. If G_k is a gamma variate with unit scale parameter and shape parameter k, then a Poisson random variable with mean $\mu = pG_k/q$, has the required negative binomial distribution. Ahrens and Dieter (1974) implemented this (NG) in FORTRAN. The prime requirement is for a Poisson generator which is efficient when μ is reset between calls. The logistic method, combined with unstored inversion for low μ, is suitable in this respect. Coupled with a good gamma generator, such as

Algorithm NB

Input $[k, p]$ Parameter $[H = 0.7]$
Output $[X]$

1    $q := 1 - p,\ m := k.$
2    $X := 0.$
3   If $p > H$ then

4        $X := \sum_{i=1}^{\langle m \rangle} \langle \ln R_i / \ln p \rangle$

5        $m := m - \langle m \rangle.$
6   End If.
7   generate $R \sim U(0, 1),\ j := 0,\ Y := q^m.$
8   While $Y < R$ do
9        $R := R - Y$
10       $j := j + 1$
11       $Y := Yp(j + m - 1)/j$
12   End While.
13   $X := X + j,$ Exit.

Table 5.3 Mean time to generate one negative binomial variate on a DEC-20 (μs) and a BBC microcomputer (ms) using Algorithm NB, with k and p reset (BBC times in parentheses)

k	p 0.1	0.5	0.9
0.1	145 (45)	154 (34)	169 (55)
1	44 (14)	64 (23)	243 (56)
2	151 (16)	189 (38)	230 (79)
5	169 (23)	269 (70)	451 (147)
10	230 (40)	363 (123)	781 (261)
30	239 (94)	881 (392)	2206 (718)

G4, this would give execution times ($k \geqslant 1$) of the order of 600–900 μs on a DEC 20, which is perhaps too large even for a portable method.

The recommendation, therefore, for negative binomial variates is to use Algorithm NB, except for extreme values of k and p. This is implemented as FORTRAN 77 function NBINOM in Appendix 1.14. In other cases the alias method described in the next section gives good performance, returning marginal generation times of the order of 60 μs on a DEC-20.

5.4 A family of alias algorithms for the binomial, Poisson, and negative binomial distributions

One of the benefits of the alias method described in Section 3.8 is that marginal generation time is extremely low and mainly independent of the parameters of the distribution. A disadvantage is that it can be used only on p.m.f.s having finite domain. For the Poisson (P) and negative binomial distributions (NB) we avoid this problem by sampling from the body of the distribution $(0 \leqslant X \leqslant m-1)$ with probability $\sum_0^{m-1} f_X(x)$ using an alias table, and from the tail of the distribution $(X \geqslant m)$ with residual probability using envelope rejection. The chosen value of m is not critical, and in this case is taken as approximately 2.5 standard deviations above the mean. In this way the efficiency of the tail variate generator is not crucial to the overall efficiency of the routine. No truncation at m is necessary for the binomial distribution (B), so we may formally set $m = n+1$.

Tail variate, X_T, generation uses a geometric target distribution

$$g_Y(x) = (1-\rho)\rho^{x-m} \qquad (x \geqslant m),$$

where $0 < \rho < 1$. Thus for the Poisson and negative binomial distributions

$$\frac{f_{X_T}(x)}{g_Y(x)} = \begin{cases} B(\mu/\rho)^x/x! & \text{(P)} \\ A(p/\rho)^x \Gamma(x+k)/\Gamma(x+1) & \text{(NB)}, \end{cases} \tag{5.5}$$

where $x \geqslant m$ and A and B are constants. The maxima of (5.5) occur at $x = m$ if we set

$$\rho = \begin{cases} \mu/(m+1) & \text{(P)} \\ p(m+k)/(m+1) & (k \geqslant 1) \\ p & (k < 1). \end{cases} \quad \text{(NB)}$$

This gives an acceptance condition

$$R_2 < \begin{cases} (\mu/\rho)^{Y-m}\Gamma(m+1)/\Gamma(Y+1) & \text{(P)} \\ (p/\rho)^{Y-m}\Gamma(Y+k)\Gamma(m+1)/\{\Gamma(m+k)\Gamma(Y+1)\}. & \text{(NB)} \end{cases}$$

Algorithm DISFAM gives details of a generator using this hybrid alias/tail approach.

Algorithm DISFAM

$$\text{S1} \quad m := \begin{cases} (n+1) & \text{(B)} \\ 1 + \langle \mu + 2.5\sqrt{\mu} \rangle & \text{(P)} \\ 1 + \langle kp/q + 2.5\sqrt{(kp)/q} \rangle & \text{(NB)} \end{cases}$$

Algorithm DISFAM (Continued)

$$S2 \quad \rho := \begin{cases} \infty & \text{(B)} \\ \mu/(m+1) & \text{(P)} \\ p(m+k)/(m+1) & (k \geqslant 1) \\ p & (k < 1) \end{cases} \quad \text{(NB)}$$

$$S3 \quad g := \sum_0^{m-1} f_X(x)$$

S4   For $x := 0, \ldots, m-1$

$$S5 \quad h_X(x) := \frac{f_X(x)}{g}$$

S6   Next x.

S7   Call Alias subroutine [Input $\{h_X(x): x=0, \ldots, m-1\}$, Output $\{$cut-off and alias values $c_x, a_x: x=0, \ldots, m-1\}$].

1   generate $R_1 \sim U(0, 1)$.

2   If $R_1 \leqslant g$ then

3          $X := mR_1/g$

4   If $X - \langle X \rangle \leqslant c_{\langle X \rangle}$ then

5              $X := \langle X \rangle$

6   Else

7              $X := a_{\langle X \rangle}$

8   End If.

9   Else

10   generate E, a standard negative exponential variate.

11         $X := m + \langle -E/\ln \rho \rangle$.

12   If $X = m$ exit.

13   generate $R_2 \sim U(0, 1)$.

14   If $R_2 > \prod_{j=1}^{X-m} B_j$

$$15 \quad \text{where } B_j := \begin{cases} \dfrac{\mu}{(X+1-j)\rho} & \text{(P)} \\ \dfrac{(X+k-j)p}{(X+1-j)\rho} & \text{(NB)} \end{cases}$$

16   goto 10.

17   End If.

18   Exit.

For ease of exposition the specification of input, output, and saved variables and functions has been relaxed. The parameter ρ is set to a very large number for the binomial distribution, although control would never be passed to line 9 for the binomial, as $g = 1$. Steps 9–16 provide the tail variates, when required. Since $B_j \leqslant 1 (1 \leqslant j \leqslant X - m)$ rejection can be ascertained whenever a partial product in line 14 becomes smaller than R_2.

5.5   Some monotonically decreasing probability mass functions

It is sometimes necessary to model situations where the probability mass function is monotonically decreasing. Included in this category are the geometric, logarithmic series, zeta, yule and harmonic distributions. We shall consider the first three only. Theoretical properties and applications of all such distributions may be found in Johnson and Kotz (1969).

Consider a negative binomial distribution

$$f_X(x) = \frac{\Gamma(x+k)p^x q^k}{\Gamma(k)\Gamma(x+1)} \qquad (x = 0, 1, \ldots), \tag{5.6}$$

where $k > 0$. If $k = 1$ this reduces to

$$f_X(x) = p^x q \qquad (x = 0, 1, \ldots).$$

Thus $X + 1$ is a geometric variate and a convenient generation method is inversion as described in Section 3.1. If p is small then it is quicker to use unstored inversion [Algorithm 3.11 with $g_j(r) = pr$]. This eliminates the time-consuming logarithmic evaluation.

As $k \to 0$ in eqn (5.6), the distribution coalesces onto $x = 0$. However, if $x = 0$ is excluded, the remaining limiting probabilities are proportional to p^x/x, giving the logarithmic series distribution,

$$f_X(x) = -p^x/(x \ln q). \qquad (x = 1, 2, \ldots). \tag{5.7}$$

This has been used to model data in biological and purchasing situations, as in Fisher *et al.* (1943), Rowe (1942), and Chatfield *et al.* (1966). From eqn (5.7), $f_X(x+1)/f_X(x) = px/(x+1)$, showing that the probabilities decay more rapidly than in the corresponding geometric distribution. As in the geometric, a long tail occurs if p is large.

Kemp (1981) proposed two strategies for generating logarithmic variates. One is based on unstored inversion [with $g_j(r) = rpj/(j+1)$]. This is recommended when $p \lesssim 0.90$. For larger values, $E(X)$ is large and an alternative is preferable. This uses the property that if Y has c.d.f.

$$F_Y(y) = \frac{\ln(1-y)}{\ln q} \qquad (0 \leqslant y \leqslant p) \tag{5.8}$$

and X has p.m.f.

$$f_X(x) = (1 - Y) Y^{x-1} \qquad (x = 1, 2, \ldots), \qquad (5.9)$$

then the unconditional distribution of X is logarithmic. Given random numbers R_1 and R_2, inversion on (5.8) and (5.9) yields

$$Y = 1 - q^{R_1}$$

and

$$X = \langle 1 + (\ln R_2 / \ln Y) \rangle,$$

respectively. For values of p not exceeding 0.999, the probability that the delivered value is either 1 or 2 is at least 0.21. Kemp notes, therefore, that it is beneficial to incorporate pretests reducing the number of logarithmic evaluations.

Turning now to the zeta distribution, this has p.m.f.

$$f_X(x) = A x^{-(\rho+1)} \qquad (x = 1, 2, \ldots) \qquad (5.10)$$

where $\rho > 0$, $A = \{\zeta(1+\rho)\}^{-1}$ and $\zeta(\cdot)$ denotes the Riemann zeta function. As ρ increases, the distribution becomes shorter tailed with more of the probability concentrated at $x = 1$. The distribution has been used in such diverse areas as linguistics (Zipf, 1949; Parunak, 1979), insurance (Seal, 1947), and modelling rare events (Bendell and Samson, 1981). Theoretical properties are to be found in Hill (1974), Hill and Woodroofe (1975a, b), and Chen (1980). There is no guidance in the literature on how to generate variates from this distribution, yet the diversity of applications suggests that simulation methods might be useful. In the form (5.10) the distribution is a one-parameter family. It seems sensible to introduce some flexibility in the shapes offered and we will consider a generalization to a two-parameter zeta distribution,

$$f_X(x) = B(x + k)^{-(\rho+1)} \qquad (x = 1, 2, \ldots), \qquad (5.11)$$

where $\rho > 0$, $k \geq 0$, and B is a constant. When k is integral this is merely a truncated one-parameter distribution, translated by k units. When k is non-integral, intermediate cases result. For $k > 0$, the two-parameter distribution has relatively less probability concentrated at $x = 1$ (compared with $k = 0$). From (5.11)

$$E(X) = \sum_{x=1}^{\infty} \frac{Bx}{(x+k)^{\rho+1}}.$$

For integral k this gives

$$E(X) = \frac{\zeta(\rho) - \sum_{1}^{k} j^{-\rho}}{\zeta(1+\rho) - \sum_{1}^{k} j^{-(\rho+1)}} - k,$$

showing that the mean is infinite whenever $\rho \leq 1$. For $\rho > 1$ and not too close to 1, $E(X)$ is not too large, and generation via inversion is a possibility. A disadvantage of this approach is that the constant $B^{-1} = \sum_1^\infty (x+k)^{-(\rho+1)}$ has to be calculated. This is time consuming, and requires care since the series converges slowly. Therefore there is advantage in developing an envelope rejection method, where the p.m.f. (5.11) need be specified only up to a multiplicative constant.

To generate discrete two-parameter zeta variates it suffices to generate continuous variates X having p.d.f.

$$f_X(x) = B\{\langle x + \tfrac{1}{2} \rangle + k\}^{-(\rho+1)} \qquad (x > \tfrac{1}{2})$$

and then deliver $N = \langle X + \tfrac{1}{2} \rangle$. We use a two-parameter Pareto target distribution having density

$$g_Y(x) = \rho(c + \tfrac{1}{2})^\rho (c+x)^{-(\rho+1)} \qquad (x > \tfrac{1}{2}, c > -\tfrac{1}{2}) \qquad (5.12)$$

and c.d.f.

$$G_Y(x) = 1 - \left(\frac{c+x}{c+\tfrac{1}{2}} \right)^{-\rho}.$$

Using inversion, a prospective variate is generated via

$$Y = (c + 0.5) R_1^{-1/\rho} - c.$$

From (5.11) and (5.12)

$$\frac{f_X(x)}{g_Y(x)} = \frac{B(x+c)^{\rho+1}}{\{\langle x + \tfrac{1}{2} \rangle + k\}^{\rho+1} \rho(c+\tfrac{1}{2})^\rho}. \qquad (5.13)$$

The maximum of (5.13) must occur at $x = n + 0.5 - \varepsilon$ where $\varepsilon \to 0+$ and n is some positive integer. To determine the M-value we must therefore maximize (w.r.t. n)

$$T(n) = \frac{n + 0.5 + c}{n + k}.$$

If $0.5 + c \geq k$, $T(n)$ is maximized at $n = 1$. Otherwise the maximum is attained as $n \to \infty$. Thus

$$M(c) = \max_x [f_X(x)/g_Y(x)]$$

$$= \begin{cases} \dfrac{B}{\rho(c+\tfrac{1}{2})^\rho} & (c < k - 0.5) \\[4mm] \dfrac{B}{\rho(c+\tfrac{1}{2})^\rho} \left(\dfrac{1.5+c}{1+k} \right)^{\rho+1} & (c \geq k - 0.5). \end{cases}$$

c is optimized by noting that the first branch of $M(c)$ is minimized at $k-0.5$. The second branch is minimized at $k-0.5$ ($\rho<k$) or $\rho-0.5$ ($\rho\geqslant k$). Given the continuity of $M(c)$ at $c=k-0.5$, the sampling efficiency is therefore maximized when

$$c=\begin{cases} k-0.5 & (\rho<k) \\ \rho-0.5 & (\rho\geqslant k), \end{cases}$$

and the associated reciprocal sampling efficiencies are

$$M=\begin{cases} \dfrac{B}{\rho k^\rho} & (\rho<k) \\[2ex] B\left\{\dfrac{1+\rho}{\rho(1+k)}\right\}^{\rho+1} & (\rho\geqslant k). \end{cases} \qquad (5.14)$$

M remains bounded for all values of ρ and k. Specimen sampling efficiencies M^{-1} are given in Table 5.4. These indicate robustness of the method over a wide range of parameter values. In particular when $k=0$ (the one-parameter zeta) the asymptotic sampling efficiency is e^{-1}.

Table 5.4 Sampling efficiencies for two-parameter zeta rejection method

k	ρ				
	$0+$	0.5	1	2	5
0	1.00	0.50	0.41	0.36	0.34
1	1.00	0.80	0.64	0.48	0.37
2	1.00	0.88	0.79	0.62	0.42
3	1.00	0.91	0.85	0.72	0.47

From eqns (5.13) and (5.14) the acceptance condition becomes

$$R_2<\begin{cases} \left\{\dfrac{X+c}{\langle X+\frac{1}{2}\rangle+k}\right\}^{\rho+1} & (\rho<k) \\[2ex] \left\{\dfrac{X+c}{\langle X+\frac{1}{2}\rangle+k}\right\}^{\rho+1}\left(\dfrac{1+k}{1+\rho}\right)^{\rho+1} & (\rho\geqslant k). \end{cases}$$

Algorithm Z gives details of the resulting generator. Note than when $\rho<1$, integer overflow of N is possible, when R_1 is small. This is likely to occur only

Algorithm Z

Input $[\rho, k]$ Saved $[c, D]$
Output $[N]$

S1   If $\rho < k$ then
S2 $c := k - 0.5,\ D := 0.$
S3   Else

S4 $c := \rho - 0.5,\ D := (1 + \rho) \ln\left(\dfrac{1+k}{1+\rho}\right).$

S5   End If.
1   generate $R_1, R_2 \sim U(0, 1).$
2     $X := (c + 0.5) R_1^{-1/\rho} - c.$
3     $N := \langle X + 0.5 \rangle.$
4     $E := -\ln R_2.$

5   If $E \leqslant (1 + \rho) \ln\left(\dfrac{N+k}{X+c}\right) - D$ goto 1.

6   Exit.

for extremely long-tailed distributions ($\rho \lesssim 0.5$). In the event of integer overflow, X and N are indistinguishable within integer machine precision, and so step 5 simply reduces to $E > -D$. If this is satisfied, the variate is simply delivered as the real quantity X.

6
Multivariate distributions

Most of the developments in variate generation have been concerned with univariate distributions. Little guidance exists for the experimenter who wishes to generate random vectors from multivariate distributions. Yet most simulations will contain some element of dependency. For example, in simulating the operations of a port it is likely that the unloading time and size of ships are statistically dependent random variables. In neutron physics the emergent energy of a particle and its direction may be dependent. Risk analysis is one application area in which a concerted effort has been made to recognize and explicitly model dependency between random variables. Thus in a problem where we wish to investigate the profitability of a new product over its entire life-cycle, important random variables might include the fixed costs, variable costs, selling price, initial investment, life of product, and share of market, which may be considered to be statistically dependent. Risk analysis simulations are described in Eilon and Fowkes (1973), Hull (1977), and Kotas and Lau (1978). However, the techniques developed for such applications are either table-based methods for discrete distributions, or a regression approach for two statistically associated random variables. No general methodology exists for standard multivariate distributions. If such methods were more widely available, it is likely that simulation would be used more extensively in economics, in modelling reliability structures, and as an investigatory tool in stochastic processes. In this chapter we describe four general approaches to the problem, together with methods for some specific distributions.

6.1 Some general methods

6.1.1 Method of conditional distributions

Suppose the p.d.f. of a random vector X is $f_X(\cdot)$, where $X' = (X_1, X_2, \ldots, X_n)$. Then this method generates the following random variables in sequence:

$$X_1; X_2|X_1; X_3|X_2, X_1; \ldots, X_n|X_{n-1}, X_{n-2}, \ldots, X_1.$$

The success of the method depends upon the ease with which expressions for the conditional distributions may be obtained, and the speed at which variates may be sampled from them.

6.1.2 Transformation to independent form

In certain cases it may be possible to obtain a transformation $Y = h(X)$ such that the components of Y are independent random variables. The practicability of such a method depends on whether variates can easily be sampled from the marginal distributions of Y and whether the inverse $X = h^{-1}(Y)$ is readily obtainable.

6.1.3 Rejection method

A straightforward generalization of the univariate envelope rejection method gives Algorithm 6.1. The choice of a target distribution $g_Y(\cdot)$ is governed by two considerations. Firstly, it should be easy to sample from, and since it is the dependence between the components of X which usually precludes direct generation, it seems reasonable to select Y in such a way that the components are independently distributed. Secondly, it is preferable if $g_Y(\cdot)$ imitates the function $f_X(\cdot)$ as far as is practicable. These requirements suggest that it many cases a suitable choice is one obtained by considering the product of the marginal densities of X, that is,

$$g_Y(\cdot) = \prod_{i=1}^{n} f_{X_i}(\cdot).$$

In the limiting case of independent components for Y, this choice ensures that all variates from the target distribution are accepted.

Algorithm 6.1

Input $[f_X(\cdot), g_Y(\cdot)]$ Saved $[M]$
Output $[X]$

S1 $M := \underset{x}{\text{Max}} [f_X(x)/g_Y(x)]$

1   generate Y from $g_Y(\cdot)$ [first stage sampling]
          and $R \sim U(0, 1)$.
2   If $R > f_X(Y)/\{Mg_Y(Y)\}$ goto 1.
3     $X := Y$, Exit.

6.1.4 Stochastic model methods

As in the case of univariate distributions, it is sometimes possible to simulate some process where a statistic can be constructed having precisely the required distribution. Since many multivariate distributions have been devised specifically to model certain physical phenomena, this approach can be expected to yield easily understood generation methods in a number of cases.

6.2 Multivariate normal distribution

This is the best known of the multivariate distributions. A row vector $X' = (X_1, \ldots, X_n)$ is distributed as multivariate normal ($X \sim N(\mu, V)$) if its p.d.f. is

$$f_X(x) = (2\pi)^{-n/2} |V|^{-\frac{1}{2}} \exp\{-\tfrac{1}{2}(x-\mu)' V^{-1}(x-\mu)\}$$

where V (the variance–covariance matrix) is any positive definite symmetric matrix. The method of transformation to independent form works well. Consider a transformation $X = cZ + \mu$, where c is the unique lower triangular matrix satisfying $cc' = V$. Then the n components of Z are distributed independently as $N(0, 1)$ random variables. To obtain the matrix c, the Cholesky decomposition is used. This enables the elements of c to be obtained recursively via

$$c_{ii} = \sqrt{\left(V_{ii} - \sum_{m=1}^{i-1} c_{im}^2 \right)}$$

$$c_{ji} = \left\{ V_{ij} - \sum_{m=1}^{i-1} c_{im} c_{jm} \right\} \Big/ c_{ii} \qquad (j > i)$$

for $i = 1, \ldots, n$, where a zero upper summation limit indicates that the sum is zero. This forms the set-up part of the complete variate generator, details of which are shown in Algorithm MVN. Steps S1–S10 perform the Cholesky decomposition. A convenient way to implement this is to store the upper triangular part of V in a vector D where

$$V_{ij} = D_{i+j(j-1)/2} \qquad (i \leqslant j \leqslant n)$$

and the lower triangular part of c in a vector F where

$$c_{ij} = F_{i+(j-1)(2n-j)/2} \qquad (1 \leqslant j \leqslant i)$$

Steps 1–10 apply the transformation $X = \mu + cZ$. Note that intermediate storage of Z is not required. Both the set-up and marginal generation stages have a number of computations which vary as n^2, indicating how both times will vary with the size of the matrix. A FORTRAN subroutine MVNORM appears in Appendix 1.15. This calls the polar Box–Müller univariate generator RND.

The Cholesky decomposition method yields an algorithm which is both simple to implement and usually faster (see, for example, Barr and Slezak,

Algorithm MVN

Input $[n, \mu_1, \ldots, \mu_n, V_{ij}(n \geq j \geq i \geq 1)]$ Saved $[c_{ij}(n \geq i \geq j \geq 1)]$
Output $[X_1, \ldots, X_n]$

S1 $c_{11} := \sqrt{V_{11}}$.
S2 For $j := 2, n$
S3 $c_{j1} := V_{1j}/c_{11}$
S4 Next j.
S5 For $i := 2, n$

S6 $c_{ii} := \sqrt{\left(V_{ii} - \sum_{m=1}^{i-1} c_{im}^2 \right)}$

S7   For $j := i+1, n$

S8 $c_{ji} := \left(V_{ij} - \sum_{m=1}^{i-1} c_{im} c_{jm} \right) \Big/ c_{ii}$

S9   Next j.
S10 Next i.
1   For $i := 1, n$
2     $X_i := \mu_i$
3   Next i.
4   For $j := 1, n$
5   generate $Z_j \sim N(0, 1)$
6   For $i := j, n$
7       $X_i := X_i + c_{ij} Z_j$
8   Next i
9   Next j.
10 Exit.

1972) than a method based on conditional distributions. To motivate a method based on conditional distributions denote

$$X' = (X_1', X_2') \text{ where } X_1' = (X_1, \ldots, X_j) \text{ and}$$

$$X_2' = (X_{j+1}, \ldots, X_n),$$

$$\mu' = E(X'), \quad \mu_1' = E(X_1'), \quad \mu_2' = E(X_2')$$

and

$$V^{-1} = \begin{pmatrix} A_{11} & A_{12} \\ A_{21} & A_{22} \end{pmatrix}$$

where partitioning is at the jth row and column. The distribution of X_2 given X_1 is itself multivariate Normal with mean $\mu_2' - (X_1 - \mu_1)' A_{12} A_{22}^{-1}$ and variance–covariance matrix A_{22}^{-1} (see, for example, Johnson and Kotz, 1972, p. 41). Hence, given $X_1, \ldots, X_j$ the next variate X_{j+1} is delivered as

$$X_{j+1} = \mu_{j+1} - \sum_{i=1}^{j} (X_i - \mu_i)(A_{12} A_{22}^{-1})_{i1} + \sqrt{(A_{22}^{-1})_{11}} Z_{j+1}, \qquad (6.1)$$

where $Z_{j+1} \sim N(0, 1)$. In practice the evaluation of $X = \mu + cZ$ is very much simpler than use of the recursion 6.1, while the only preliminary work required is computation of c.

A third possibility is use of the rejection method with a target distribution,

$$g_Y(x) = (2\pi)^{-n/2} |\boldsymbol{\beta}|^{-\frac{1}{2}} \exp\{-\tfrac{1}{2}(x - \mu)' \boldsymbol{\beta}^{-1}(x - \mu)\}$$

where

$$\boldsymbol{\beta} = \begin{pmatrix} \beta_1 & & & \\ & \beta_2 & & \\ & & \ddots & \\ & & & \beta_n \end{pmatrix}$$

giving

$$\frac{f_X(x)}{g_Y(x)} = \frac{|\boldsymbol{\beta}|^{\frac{1}{2}}}{|V|^{\frac{1}{2}}} \exp\{-\tfrac{1}{2}(x - \mu)'(V^{-1} - \boldsymbol{\beta}^{-1})(x - \mu)\}.$$

$\boldsymbol{\beta}$ is selected so that $V^{-1} - \boldsymbol{\beta}^{-1}$ is positive definite. This is achieved by ensuring that every diagonal element of $\boldsymbol{\beta}$ is no smaller than the largest eigenvalue of V. In this case $M = \sqrt{\{|\boldsymbol{\beta}|/|V|\}}$ showing that the optimal choice of $\boldsymbol{\beta}$ is to set

$$\beta_i = \max_{1 \leqslant j \leqslant n} \{\lambda_j\}$$

for all i, where $\{\lambda_j\}$ are the eigenvalues of V. The acceptance condition becomes

$$R < \exp\{-\tfrac{1}{2}(Y - \mu)'(V^{-1} - \boldsymbol{\beta}^{-1})(Y - \mu)\}$$

or

$$E > \tfrac{1}{2}(Y - \mu)'(V^{-1} - \boldsymbol{\beta}^{-1})(Y - \mu). \tag{6.2}$$

There are several disadvantages to such a method. Firstly, the sampling efficiency approaches zero as $|V| \to 0$. Secondly, finding the largest eigenvalue routinely poses some problems. Lastly, calculation of the right-hand side of the acceptance condition (6.2) involves more work than evaluation of cZ in the Cholesky decomposition method.

6.3   Wishart distribution

Suppose $x_1, \ldots, x_N$ are N independent vectors drawn from a p-variate normal distribution, $N(0, V)$. Forming the $N \times p$ data matrix X, where the ith row of X is x_i, the maximum likelihood estimator of V is the sample covariance $S = X'X/N$. The $p \times p$ matrix $B = NS$ is said to have a Wishart distribution with scale parameter V and N degrees of freedom. The domain of variation of the distribution function is over all $p \times p$ positive definite non-singular matrices $\boldsymbol{\beta}$. For non-singularity it is necessary that $N \geqslant p$, that is the sample size should be at least as large as the dimension of the multivariate normal distribution.

A direct, if inefficient generation method is to deliver $\boldsymbol{B} = \boldsymbol{X}'\boldsymbol{X}$, requiring Np standard normal deviates and $O(Np^2)$ product operations. Smith and Hocking (1972) proposed the following alternative method. To sample from a Wishart distribution with parameters N and $\boldsymbol{I}$ deliver $\boldsymbol{B} = \boldsymbol{A}'\boldsymbol{A}$ where $\boldsymbol{A}$ is an upper triangular $p \times p$ matrix, in which the ith diagonal element is distributed as $\chi^2_{N-i+1}(i=1, \ldots, p)$ and the off-diagonal elements are distributed as standard normal deviates. To sample from a Wishart distribution with parameters N and $\boldsymbol{V}$, $\boldsymbol{B} = \boldsymbol{c}'\boldsymbol{A}'\boldsymbol{A}\boldsymbol{c}$ is delivered where $\boldsymbol{c}$ is the lower triangular matrix in a Cholesky decomposition of $\boldsymbol{V}$. Algorithm WISH shows in principle how random $p \times p$ Wishart matrices with parameters N and $\boldsymbol{V}$ may be generated. Such a method requires $p(p-1)/2$ standard normal deviates and p chi-squared variates, a total of $p(p+1)/2$ variates. Since $N \geqslant p$, this is less than the total required using the direct method, although generation of a chi-squared variate will take a little longer than a normal deviate. The number of product operations is $O(p^3)$ compared with $O(N^2 p)$ in the direct method. Unless N is close to p and chi-squared variate generation extremely slow relative to normal generation, Algorithm WISH can be expected to be faster than the direct method. Smith and Hocking used a Wilson–Hilferty approximation to generate chi-squared variates, and consequently recommend their procedure only when $N \geqslant 20$. Nowadays efficient chi-squared (gamma) generators are available and so the restriction on N becomes redundant.

Algorithm WISH

Input $[\boldsymbol{V}]$ Saved $[\boldsymbol{c}]$
Output $[\boldsymbol{B}]$

S1   Construct $\boldsymbol{c}$ such that $\boldsymbol{cc}' = \boldsymbol{V}$.

$$1 \quad A_{ij} := \begin{cases} Z & (1 \leqslant i < j \leqslant p) \\ \chi^2_{N-i+1} & (1 \leqslant i = j \leqslant p) \\ 0 & (i > j) \end{cases}$$

2    $\boldsymbol{H} := \boldsymbol{Ac}$.
3    $\boldsymbol{B} := \boldsymbol{H}'\boldsymbol{H}$.
4   Exit.

6.4   Bivariate exponential and Poisson distributions

Any bivariate distribution having negative exponential marginals may be termed a bivariate exponential distribution. Given the duality between the exponential and Poisson distributions, it is not surprising to find bivariate Poisson distributions constructed in a similar manner. In this section we discuss generation methods for both types of distribution, concentrating mainly on those that arise in a reliability context.

6.4.1.   A memoryless bivariate exponential distribution

Consider a two-component system with lifetimes T_1 and T_2, respectively. The system is subjected to 'fatal' shocks according to three independent Poisson processes, rates λ_1, λ_2, λ_{12}, respectively. The first process applies shocks to component 1 only, the second to component 2 only, and the third simultaneously to both components. Denoting the probability that components 1 and 2 survive to t_1 and t_2, respectively, by $R_{T_1, T_2}(t_1, t_2)$, we have

$$R_{T_1, T_2}(t_1, t_2) = \prod_{j=1}^{3} \text{prob(zero shocks in time } t_j \text{ from process } j)$$

where $t_3 = \max(t_1, t_2)$. Thus

$$R_{T_1, T_2}(t_1, t_2) = \exp\{-\lambda_1 t_1 - \lambda_2 t_2 - \lambda_{12} \max(t_1, t_2)\} \qquad (6.3)$$

showing that the marginal survivor functions are

$$R_{T_i}(t_i) = e^{-(\lambda_i + \lambda_{12})t_i} \qquad (i = 1, 2).$$

Thus the marginal lifetimes are negative exponential with means $(\lambda_i + \lambda_{12})^{-1}$ ($i = 1, 2$). The importance of this distribution (Marshall and Olkin, 1967) is that it is the only bivariate exponential distribution satisfying the 'memoryless' property

$$R_{T_1, T_2}(s_1 + t, s_2 + t) = R_{T_1, T_2}(t, t) R_{T_1, T_2}(s_1, s_2)$$

for $s_1, s_2, t \geq 0$. That is, the survivor function of a pair of components of *common* age is identical to a pair of new components. The distribution (6.3) is a mixture of discrete and continuous with density,

$$f_{T_1, T_2}(t_1, t_2) = \begin{cases} \lambda_2(\lambda_1 + \lambda_{12})e^{-(\lambda_1 t_1 + \lambda_2 t_2 + \lambda_{12} t_1)} & (0 \leq t_2 < t_1) \\ \lambda_1(\lambda_2 + \lambda_{12})e^{-(\lambda_1 t_1 + \lambda_2 t_2 + \lambda_{12} t_2)} & (0 \leq t_1 < t_2) \end{cases} \qquad (6.4)$$

and

$$\text{prob}(T_1 = T_2) = \lambda_{12}/(\lambda_1 + \lambda_2 + \lambda_{12}),$$

the discrete part arising from the possibility of simultaneous failure of the components.

The genesis of the distribution suggests a stochastic model approach to variate generation. Denoting the time to the first event in each of the three Poisson processes by independent negative exponential variates Y_1, Y_2, Y_{12} with means λ_1^{-1}, λ_2^{-1}, λ_{12}^{-1}, we deliver $T_1 = \min(Y_1, Y_{12})$ and $T_2 = \min(Y_2, Y_{12})$.

Alternatively, the method of conditional distributions may be used. From eqn (6.4) the conditional density of T_2 given $T_1 = t$ is

$$f_{T_2 | T_1}(t_1, t_2) = \begin{cases} \lambda_2 e^{-\lambda_2 t_2} & (t_2 < t_1) \\ \dfrac{\lambda_1(\lambda_2 + \lambda_{12}) e^{-(\lambda_2 t_2 + \lambda_{12} t_2 - \lambda_{12} t_1)}}{\lambda_1 + \lambda_{12}} & (t_1 < t_2). \end{cases}$$

Hence the conditional distribution function is

$$F_{T_2 | T_1}(t_1, t_2) = \begin{cases} 1 - e^{-\lambda_2 t_2} & (t_2 < t_1) \\ 1 - \dfrac{\lambda_1 e^{\lambda_{12} t_1} e^{-(\lambda_2 + \lambda_{12}) t_2}}{\lambda_1 + \lambda_{12}} & (t_1 < t_2). \end{cases}$$

The discreteness arises from the discontinuity at $t_2 = t_1$ giving

$$\text{prob}(T_2 = t_1 | T_1 = t_1) = \frac{\lambda_{12} e^{-\lambda_2 t_1}}{\lambda_1 + \lambda_{12}}.$$

Algorithm BEXP generates T_1 from a negative exponential, mean $(\lambda_1 + \lambda_{12})^{-1}$ and then T_2, using inversion on the conditional distribution.

Algorithm BEXP

Input $[\lambda_1, \lambda_2, \lambda_{12}]$
Output $[T_1, T_2]$.

1   generate E_1, $E_2 \sim$ standard negative exponential variates.
2   $T_1 := E_1/(\lambda_1 + \lambda_{12})$, $T_2 := E_2/\lambda_2$.
3   If $T_2 \geqslant T_1$ then

4       $\Delta := T_2 - T_1 - \dfrac{1}{\lambda_2} \ln\left[\dfrac{\lambda_1 + \lambda_{12}}{\lambda_1} \right]$

5   If $\Delta < 0$ then
6           $T_2 := T_1$
7   Else

8           $T_2 := T_1 + \dfrac{\lambda_2 \Delta}{\lambda_2 + \lambda_{12}}$

9   End If.
10 End If.
11 Exit.

The conditional distribution method requires little extra programming over the stochastic model method. Only two exponential variates are needed per vector, compared with three using the stochastic model method. This offers a useful saving in time. Programmed in BASIC on a BBC micro-computer, the two methods gave marginal generation times of approximately 48 and 66 ms per vector, independently of parameter values. We may learn from this that although stochastic model methods are often intuitively appealing, other methods may be faster and not appreciably harder to implement. It is interesting to note that Y_1, Y_2, and Y_{12} are independent random variables, so the stochastic model method may also be interpreted as a transformation to independent form method.

The bivariate exponential may be extended to an n-variate distribution. In this case Marshall and Olkin consider n components. There are n inde-pendent Poisson processes, rate $\{\lambda_i\}$, giving shocks to components $1, \ldots, n$, respectively. A further $\binom{n}{2}$ independent processes with rates λ_{ij} $(i<j)$ give shocks simultaneously to components i and j. In general, there is a group of $\binom{n}{m}$ independent processes $(m=1, \ldots, n)$ each giving shocks to exactly m of the n components. Thus the system is subjected to shocks from a total of $\sum_1^n \binom{n}{m}=2^n-1$ processes. This is the required number of exponential variates needed to generate a vector $(T_1, \ldots, T_n)$ using a stochastic model method. For example, if $n=3$, there are seven processes with rates $\lambda_1, \lambda_2, \lambda_3, \lambda_{12}, \lambda_{13}, \lambda_{23}, \lambda_{123}$. Denoting, $Y., Y.., Y...$ as the associated exponential variates, we have

$$T_1 = \min(Y_1, Y_{12}, Y_{13}, Y_{123})$$
$$T_2 = \min(Y_2, Y_{12}, Y_{23}, Y_{123})$$
$$T_3 = \min(Y_3, Y_{13}, Y_{23}, Y_{123}).$$

Clearly the marginal generation time for vectors T increases exponentially with the dimension of the vector. A method based on conditional distri-butions would require only n random numbers (compared with 2^n-1). Unfortunately the multiple discontinuities in the conditional distributions make such an approach impracticable, except for small n. Thus for general n, the stochastic model method must suffice.

6.4.2   Bivariate Poisson distribution

Consider now the two-component system introduced in Section 6.4.1. Sup-pose that a component is replaced with an identical one immediately it receives a fatal shock. Denote the number of replacements made for compo-nents 1 and 2 during time t by $N_1(t)$ and $N_2(t)$. (Note that due to the memoryless property of this bivariate exponential, we insist only that the components be of identical age at $t=0$, and not necessarily 'new'.) The p.m.f. of $N_1(t)$ and $N_2(t)$ is

$f_{N_1, N_2}(n_1, n_2) = \text{prob}(\text{processes 1 and 3 jointly yield } n_1 \text{ shocks and processes 2 and 3 jointly yield } n_2 \text{ shocks})$

$$= \sum_{j=0}^{\min(n_1, n_2)} \frac{(\lambda_1 t)^{n_1-j} e^{-\lambda_1 t}}{(n_1-j)!} \frac{(\lambda_2 t)^{n_2-j} e^{-\lambda_2 t}}{(n_2-j)!} \frac{(\lambda_{12} t)^j e^{-\lambda_{12} t}}{j!}.$$

Without loss of generality we set $t=1$, giving

$$f_{N_1, N_2}(n_1, n_2) = e^{-(\lambda_1 + \lambda_2 + \lambda_{12})} \sum_{j=0}^{\min(n_1, n_2)} \frac{\lambda_1^{n_1-j} \lambda_2^{n_2-j} \lambda_{12}^j}{(n_1-j)!(n_2-j)!j!} \tag{6.5}$$

The marginal distributions of N_1 and N_2 are clearly Poisson, mean $\lambda_1 + \lambda_{12}$ and $\lambda_2 + \lambda_{12}$, respectively, and so the joint p.m.f. (6.5) may justifiably be called a bivariate Poisson distribution.

Kemp and Loukas (1978) discuss some generation methods. The simplest to implement is to deliver $N_1 = X_1 + X_{12}$, $N_2 = X_2 + X_{12}$, where X_1, X_2, and X_{12} are independent Poisson variates mean λ_1, λ_2, and λ_{12}, respectively. Alternatively, the conditional distribution of N_2 given $N_1 = n_1$ is

$$f_{N_2 | N_1}(n_1, n_2) = e^{-\lambda_2} \sum_{j=0}^{\min(n_1, n_2)} \binom{n_1}{j} \left(\frac{\lambda_1}{\lambda_1 + \lambda_{12}}\right)^{n_1-j} \left(\frac{\lambda_{12}}{\lambda_1 + \lambda_{12}}\right)^j \frac{\lambda_2^{n_2-j}}{(n_2-j)!}$$

which is identical to the distribution of $X_2 + B$, where X_2 is Poisson mean λ_2 and B is binomial with parameters n_1 and $\lambda_{12}/(\lambda_1 + \lambda_{12})$. To generate (N_1, N_2), sample N_1 from a Poisson distribution mean $\lambda_1 + \lambda_{12}$ and then deliver $N_2 = X_2 + B$. This conditional distribution method requires two Poisson and one binomial variates, while the stochastic model method needs three Poisson variates, so the two methods can be expected to give comparable speeds in a variety of computing environments.

Kemp and Loukas found both these methods to be relatively slow when compared with 'modified inverse methods'. These are essentially table-based conditional distribution methods. They used slow multiplicative (Poisson) and Bernoulli process (Binomial) generators for the stochastic model and conditional distribution methods. In general, the latter methods will be competitive when efficient Poisson and binomial generators are used. They are also easy to program and avoid the truncation which is a feature of table based methods.

A final possibility for this distribution is to use Algorithm BEXP when $\lambda_1 + \lambda_{12}$ and $\lambda_2 + \lambda_{12}$ are both small ($\lesssim 1$). Joint times to failure (T_1, T_2) can be generated successively until the partial sum for each component exceeds 1.

6.4.3   Other shock models leading to bivariate and multivariate exponential distributions

In modelling some reliability situations, although exponential marginals are required, it may not be appropriate to consider the Marshall–Olkin bivariate exponential because of the discontinuity at $t_1 = t_2$.

Consider first a single component which is subjected to shocks that occur according to a Poisson process, rate λ. Suppose further that the probability that a shock is 'fatal' is $(1 - \rho)$, $0 \leqslant \rho \leqslant 1$, independently of previous shocks. Then the number of shocks N till the component fails is geometrically distributed with p.m.f.

$$P(N = n) = (1 - \rho)\rho^{n-1} \quad (n \geqslant 1),$$

and the time to failure, T, has p.d.f.

$$f_T(t) = \sum_{n=1}^{\infty} \frac{(\lambda t)^{n-1} \lambda e^{-\lambda t}(1 - \rho)\rho^{n-1}}{(n-1)!}$$

$$= (1 - \rho)\lambda e^{-\lambda t(1 - \rho)},$$

that is negative exponential, with parameter $\mu = (1 - \rho)\lambda$.

Downton (1970) extends this to two components. Each component is subjected to shocks according to independent Poisson processes, rates λ_1 and λ_2, respectively. Each shock results in a common outcome for both components (either fatal or non-fatal). The physical basis for the model is rather artificial since corresponding shocks will occur at different times to the two components. Nevertheless, the joint p.d.f. of lifetimes T_1 and T_2 of the components is

$$f_{T_1, T_2}(t_1, t_2) = \sum_{n=1}^{\infty} \frac{(\lambda_1 t_1)^{n-1} \lambda_1 e^{-\lambda_1 t_1}}{(n-1)!} \frac{(\lambda_2 t_2)^{n-1} \lambda_2 e^{-\lambda_2 t_2}}{(n-1)!} (1 - \rho)\rho^{n-1}$$

$$= \lambda_1 \lambda_2 (1 - \rho) e^{-(\lambda_1 t_1 + \lambda_2 t_2)} \sum_{n=1}^{\infty} \frac{\{\sqrt{(\lambda_1 \lambda_2 t_1 t_2 \rho)}\}^{2n-2}}{\{(n-1)!\}^2}$$

$$= \lambda_1 \lambda_2 (1 - \rho) e^{-(\lambda_1 t_1 + \lambda_2 t_2)} I_0 \{2\sqrt{(\lambda_1 \lambda_2 t_1 t_2 \rho)}\},$$

where $I_0(\cdot)$ is the modified Bessel function of the first kind and order zero. The construction ensures that the marginal densities are exponential, with parameters $\mu_i = (1 - \rho)\lambda_i$, $i = 1, 2$. The joint density may be written as

$$f_{T_1, T_2}(t_1, t_2) = \frac{\mu_1 \mu_2}{1 - \rho} \exp\left\{\frac{-(\mu_1 t_1 + \mu_2 t_2)}{1 - \rho}\right\} I_0 \left\{\frac{2\sqrt{(\mu_1 \mu_2 t_1 t_2 \rho)}}{1 - \rho}\right\}, \quad (6.6)$$

where $\mu_1, \mu_2 \geqslant 0$, $0 \leqslant \rho \leqslant 1$, that is bivariate exponential with means μ_1^{-1}, μ_2^{-1}. It is a simple matter to show that the correlation between T_1 and T_2 is ρ. The distribution (6.6) is a special case of a bivariate gamma distribution considered by Krishnamoorthy and Parthasarathy (1951). How can variates T_1 and T_2 be generated from this bivariate exponential distribution? The construction suggests a stochastic model approach in which firstly the (common) number of shocks is obtained using

$$N = \left\langle 1 + \frac{\ln R}{\ln \rho} \right\rangle,$$

where $R \sim U(0.1)$. T_i is then delivered as $\text{gamma}(N)/\lambda_i = (1-\rho)$ $\text{gamma}(N)/\mu_i$, where $\text{gamma}(N)$ is a standard gamma variate with shape parameter N. An alternative method is to deliver $T_1 = \frac{1}{2}(U_1^2 + U_2^2)$ and $T_2 = \frac{1}{2}(U_3^2 + U_4^2)$ where (U_1, U_3) is bivariate normal with correlation ρ, independently of (U_2, U_4) which is also bivariate normal with correlation ρ. Al-Saadi et al. (1979) used both methods in Monte Carlo calculations of the power of tests for independence of exponential random variables, against a bivariate exponential of Downton form. In a subsequent paper, Al-Saadi and Young (1980) simulated the distribution in an investigation of the properties of various estimators of ρ. The geometric based generation method was also used by Al-Saadi and Young (1982) in an n-variate extension to the distribution. Each component receives the same number of shocks, leading to equal correlation coefficients.

A generalization of the Downton model due to Hawkes (1972) is to suppose that each shock results in a fatality to both components (with probability P_{11}), a fatality to neither (with probability P_{00}), a fatality to the first but not the second (with probability P_{10}), or a fatality to the second but not the first (with probability P_{01}). The Downton model arises as the special case $P_{10} = P_{01} = 0$, $P_{11} = 1 - \rho$, $P_{00} = \rho$. For the more general model, the number of shocks N_1 and N_2 to failure of the two components is not necessarily the same. The joint survivor function is

$$P(N_1 > n_1, N_2 > n_2) = \begin{cases} P_{00}^{n_1} Q_2^{n_2 - n_1} & (n_1 \leqslant n_2) \\ P_{00}^{n_2} Q_1^{n_1 - n_2} & (n_2 \leqslant n_1) \end{cases}$$

where $Q_1 = P_{00} + P_{01}$ and $Q_2 = P_{00} + P_{10}$. Thus

$$P(N_1 > n_1) = Q_1^{n_1} \tag{6.7}$$

and

$$P(N_2 > n_2) = Q_2^{n_2}$$

indicating that the marginal distributions are geometric, leading to marginal distributions of times to failure which are exponential. The resulting bivariate exponential distribution has five parameters. To generate T_1 and T_2, inversion based on eqn (6.7) gives

$$N_1 = \left\langle 1 + \frac{\ln R_1}{\ln Q_1} \right\rangle,$$

where $R_1 \sim U(0, 1)$. The conditional survivor function for N_2 is

$$P(N_2 > n_2 | N_1 = n_1) = \begin{cases} \left(\dfrac{P_{00}}{Q_1 Q_2}\right)^{n_1 - 1} \left(\dfrac{P_{10} Q_2^{n_2 - 1}}{1 - Q_1}\right) & (n_1 \leqslant n_2) \\ \left(\dfrac{P_{00}}{Q_1}\right)^{n_2} & (n_1 > n_2) \end{cases} \qquad (6.8)$$

Inversion on eqn (6.8) yields

$$N_2 = \begin{cases} \langle 1 + \ln R_2 / \ln(P_{00}/Q_1) \rangle \\ \langle 2 + [\ln R_2 - (N_1 - 1)\ln\{P_{00}/(Q_1 Q_2)\} + \ln(1 + P_{11}/P_{10})]/\ln Q_2] \rangle \end{cases}$$

according to whether $N_1 > \langle 1 + \ln R_2 / \ln(P_{00}/Q_1) \rangle$ or not. The final step is to deliver T_i as gamma $(N_i)/\lambda_i$, $i = 1, 2$.

6.5 Some other multivariate distributions

A good source of information on multivariate distributions is Johnston and Kotz (1969, 1972). We consider here generation methods for selected distributions only.

A bivariate beta distribution is given by

$$f_{X_1, X_2}(x_1, x_2) = \frac{\Gamma(\alpha_1 + \alpha_2 + \beta) x_1^{\alpha_1 - 1} x_2^{\alpha_2 - 1} (1 - x_1 - x_2)^{\beta - 1}}{\Gamma(\alpha_1) \Gamma(\alpha_2) \Gamma(\beta)}$$

where $\alpha_1, \alpha_2, \beta > 0$ and $x_1, x_2 \geqslant 0$ and $x_1 + x_2 \leqslant 1$. Loukas (1984) examined five generation methods. The first, motivated by Jöhnk's method for the univariate beta is to deliver $X_1 = Y_1/V$, $X_2 = Y_2/V$ subject to $V = Y_1 + Y_2 + Y_3 \leqslant 1$, where Y_i, $i = 1, 2, 3$ are independent beta variates with parameters α_1 and 1, α_2 and 1, β and 1, respectively. The second method, valid when $\beta \geqslant 1$ is to use envelope rejection with a uniform target distribution to deliver $X_1^{\alpha_1}$ and $X_2^{\alpha_2}$. When $\alpha_1, \alpha_2, \beta \geqslant 1$ envelope rejection with a uniform target distribution is used to generate X_1 and X_2 directly. The fourth approach uses the method of conditional distributions. The last method takes three independent standard gamma variates Y_1, Y_2, and Y_3 with shape parameters α_1, α_2, and β, respectively, and delivers $X_1 = Y_1/(Y_1 + Y_2 + Y_3)$ and $X_2 = Y_2/(Y_1 + Y_2 + Y_3)$. Based on timing experiments Loucas recommends the gamma method for sampling for most values of α_1, α_2, and β.

In much simulation work there is a need for non-negative random vectors, and a starting point for this is to consider a multivariate gamma distribution.

Ronning (1977) considers one such distribution constructed in the following way. Let $G_1' = (G_{\alpha_1}, G_{\alpha_2}, \ldots, G_{\alpha_k})$ and $G_2' = (G_{\beta_1}, G_{\beta_2}, \ldots, G_{\beta_N})$ be k and $N (\geqslant k)$ dimensional vectors, respectively, where $\{G_{\alpha_i}\}$ and $\{G_{\beta_i}\}$ are independent random variates drawn from standard gamma distributions with shape parameters $\{\alpha_i\}$ and $\{\beta_i\}$, respectively. Let T be a $k \times N$ incidence matrix consisting of 0s and 1s. A random vector V is defined by

$$V = G_1 + TG_2. \tag{6.9}$$

Since each element of V is a sum of independent gamma random variables, the marginal distributions of $\{V_i\}$ are also gamma. For this multivariate gamma distribution, Ronning examines various patterns for T which enable certain correlation structures to be attained. Simulation of the vector V follows directly from the construction (6.9). The method is not amenable to situations where negative correlations are required.

Turning now to two discrete distributions, the multinomial probability mass function is defined by

$$f_{X_1, \ldots, X_m}(x_1, \ldots, x_m) = N! \prod_{i=1}^{m} \frac{p_i^{x_i}}{x_i!}$$

where $0 \leqslant p_i \leqslant 1$, $\sum p_i = 1$, $N = \sum x_i$, and $0 \leqslant x_i \leqslant N$. This is an extension of the binomial distribution, with N Bernoulli trials, each trial resulting in outcome i with probability p_i. Clearly, the marginal distribution of N_i is binomial, with parameters N and p_i. Thus one convenient generation method is to generate X_1 from such a binomial distribution, X_2 from a binomial distribution with parameters $N - X_1$ and $p_2/(1 - p_1)$, X_3 from a binomial distribution with parameters $N - X_1 - X_2$ and $p_3/(1 - p_1 - p_2)$, and so on. Alternatively, the N Bernoulli trials can be simulated directly. For $N \gg m$, the former method, when used with an efficient binomial generator, is likely to be faster.

An extension of the negative binomial distribution is the negative multinomial, constructed as follows. Let p_i, $i = 1, \ldots, m+1$, denote the probability of a Bernoulli trial resulting in event i. Let X_i denote the number of occurrences of event type i until the kth occurrence of event type $m+1$. Then

$$P(X_1 = x_1, \ldots, X_m = x_m) = \frac{\Gamma\left(k + \sum_{i=1}^{m} x_i\right)(1 - p)^k \prod_{i=1}^{m} \frac{p_i^{x_i}}{x_i!}}{\Gamma(k)}$$

where $p = \sum_{i=1}^{m} p_i = 1 - p_{m+1}$. The p.m.f. is still valid for non-integer $k > 0$. The case $m = 1$ reduces to the negative binomial distribution, eqn (5.6). For integer k, simulating Bernoulli trials till the kth occurrence of event type $(m+1)$ provides a direct generation method. This requires a mean of $k/(1 - p)$ trials, so will be inefficient if k is large or p is close to 1. An alternative is to note that

$\sum\limits_{i=1}^{m} X_i$ is negative binomially distributed with parameters k and p. Conditioning on $\sum\limits_{i=1}^{m} X_i = n$, the distribution of $X_1, \ldots, X_m$ is multinomial with parameters n and p_i/p, $i = 1, \ldots, m$. This provides a relatively efficient method, when $k/(1-p)$ is large, via the generation of one negative binomial and m binomial variates.

7
Miscellaneous topics

We complete our discussion of random variate generation with a treatment of some miscellaneous topics. Section 7.1 deals with the generation of order statistics. It is seen in Section 7.2 that one method of generating events in a time inhomogeneous Poisson process is via such statistics. Also covered in Section 7.2 is the simulation of discrete state Markov processes. Section 7.3 deals with the generation of variates from distribution tails.

7.1 The generation of order statistics

Suppose a random sample of size n is drawn from a distribution having p.d.f. $f_X(\cdot)$. If the r.v.s are ranked so that $X_{(1)} \leqslant X_{(2)} \ldots \leqslant X_{(n)}$, then $X_{(j)}$ is known as the jth order statistic. There are many instances in which some or all of the order statistics need to be simulated. For example, Newby (1979) gives an example concerning the maintenance of motorway lamps, in which the time to failure for a line of lamps is governed by order statistics from the life distribution of the lamps.

A natural way to approach the problem of generating such statistics from arbitrary distributions, is to first enquire what methods are available for uniform order statistics.

7.1.1 Generation of uniform order statistics

We will denote the n order statistics from $U(0, 1)$ by $U_{(1)}, \ldots, U_{(n)}$. Three generation methods can be identified.

(i) Direct method

In this case n random numbers are drawn, and sorted in ascending order. The efficiency depends crucially on the efficiency of the sorting routine. Unsophisticated methods such as the 'Bubblesort' have execution times behaving as $O(n^2)$. Better methods include 'Quicksort' or 'Heapsort' (see, for example, Goodman and Hedetniemi, 1977, Chapter 5) which behave as $O(n \ln n)$.

A variant of this idea is to divide the interval $[0, 1)$ into k equal sub-intervals, to generate a multinomial vector (Section 6.5), $M_1, \ldots, M_k$ where $\sum M_i = n$, and to sort the M_i uniformly distributed observations within the ith interval. This 'grouping method', described by Rabinowitz and Berenson (1974), may be viewed as an attempt to make the sort more manageable, by splitting it up into a number of smaller sorts.

(ii) Sequential method

This approach utilizes the conditional c.d.f. of one order statistic, given values for either the higher or the lower order statistics. Suppose we wish to generate the order statistics in descending order. Then

$$\text{prob}(U_{(n)} < u_n) = u_n^n,$$

and

$$\text{prob}(U_{(j)} < u_j | U_{(j+1)} = u_{j+1}, \ldots, U_{(n)} = u_n)$$
$$= \text{prob}(U_{(j)} < u_j | U_{(j+1)} = u_{j+1})$$
$$= \left(\frac{u_j}{u_{j+1}}\right)^j \qquad (j = n-1, \ldots, 1).$$

Given n random numbers $R_n, \ldots, R_1$, inversion of these c.d.f.s leads to

$$U_{(n)}^n = R_n$$

and

$$\left\{\frac{U_{(j)}}{U_{(j+1)}}\right\}^j = R_j \qquad (j = n-1, \ldots, 1),$$

or

$$U_{(n)} = R_n^{1/n} \qquad (7.1)$$

and

$$U_{(j)} = U_{(j+1)} R_j^{1/j} \qquad (j = n-1, \ldots, 1).$$

Sometimes it is necessary to generate the order statistics in ascending order. This may be the case if only the lower order statistics are required. Another instance is where ascending order statistics represent event times in a simulation. For such cases we utilize the representation,

$$\text{prob}(U_{(1)} > u_1) = (1 - u_1)^n,$$

and

$$\text{prob}(U_{(j+1)} > u_{j+1} | U_{(j)} = u_j) = \left(\frac{1 - u_{j+1}}{1 - u_j}\right)^{n-j}.$$

On inversion this gives

$$1 - U_{(1)} = R_1^{1/n}$$

and

$$1 - U_{(j+1)} = \{1 - U_{(j)}\} R_{j+1}^{1/(n-j)} \qquad (j = 1, \ldots, n-1). \qquad (7.2)$$

(iii)   Exponential spacings method

This is based on the result (see, for example, Moran, 1947) that if $Z_1, \ldots,$ Z_{n+1} are i.i.d. unit negative exponential variates, then $\sum_{j=1}^{i} Z_j / \sum_{j=1}^{n+1} Z_j$ has the same distribution as $U_{(i)}$, for $i = 1, \ldots, n$.

Gerontidis and Smith (1982) performed a comparison of these three methods plus the grouping variant of the first. Their conclusions are qualified by the machine used and the method of generating exponential variates in the spacings method (they used the Forsythe–von Neumann rejection method rather than the logarithmic transformation). They found that when all the order statistics are required, the exponential spacings and grouping [using an $O(n \ln n)$ sorting algorithm] methods are comparable in speed, and superior to the others, except for small values of $n (\leqslant 20)$ when the direct method is competitive. In the grouping method the optimum number of sub-intervals was found to be of the order of $n/4$. When only the low or high order statistics are required, the exponential spacings approach may be inefficient as $(n+1)$ exponential variates still have to be generated. In these instances the sequential method may be more appropriate. For central order statistics only, Ramberg and Tadikamalla (1978) first generated the highest order statistic required. This is easily performed, since the jth order statistic follows a beta distribution with parameters j and $n - j + 1$. The remaining order statistics are then obtained using the sequential method.

7.1.2   Order statistics from general distributions

For general distributions two methods can be identified.

(i)   Direct method

In this case n random variates $X_1, \ldots, X_n$ from the parent c.d.f. $F_X(\cdot)$ are drawn. These are then sorted so that $X_{(1)} \leqslant \ldots \leqslant X_{(n)}$. Gerontidis and Smith (1982) extended the grouping idea discussed for uniform order statistics, by dividing the domain of $F_X(\cdot)$ into k intervals of equal probability. The generation of M_i random variates from the ith such interval is relatively easy if the rejection method is used, since the p.d.f. is required to be known only up to a multiplicative constant.

(ii)   Inversion method

Having obtained the uniform order statistics $\{U_{(i)}\}$ by any of the methods stated in Section 7.1.1, $\{X_{(i)}\}$ may be obtained by solving $F_X(X_{(i)}) = U_{(i)}$, or alternatively $R_X(X_{(i)}) = 1 - U_{(i)}$ where $R_X(\cdot) = 1 - F_X(\cdot)$. If the representation (7.1) is used then

$$F(X_{(n)}) = R_n^{1/n}$$

and   (7.3)

$$F_X(X_{(j)}) = F_X(X_{(j+1)}) R_j^{1/j} \qquad (j = n-1, \ldots, 1),$$

which gives a recursion for obtaining the order statistics in descending order. To obtain them in ascending order (7.2) gives

$$R_X(X_{(1)}) = R_1^{1/n}$$

and   (7.4)

$$R_X(X_{(j+1)}) = R_X(X_{(j)}) R_{j+1}^{1/(n-j)} \qquad (j = 1, \ldots, n-1).$$

Thus to generate order statistics from the negative exponential distribution, for example, $R_X(x) = e^{-x}$, indicating that generation in ascending order is more convenient. Use of (7.4) leads to

$$X_{(1)} = -\ln R_1/n$$

and

$$X_{(j+1)} = X_{(j)} - \ln R_{j+1}/(n-j),$$

or

$$X_{(1)} = E_1/n,$$

and   (7.5)

$$X_{(j+1)} = X_{(j)} + \frac{E_{j+1}}{n-j} \qquad (j = 1, \ldots, n-1),$$

where $\{E_j\}$ are standard i.i.d. negative exponential variates.

In other cases generation in descending order may be more convenient. For example, to generate order statistics from a folded Cauchy distribution,

$$f_X(x) = \frac{2}{\pi(1+x^2)} \qquad (x \geqslant 0),$$

we have

$$F_X(x) = \frac{2}{\pi} \tan^{-1} x.$$

Use of (7.3) leads to

$$\tan^{-1} X_{(n)} = \pi R_n^{1/n}/2$$

and

$$\tan^{-1} X_{(j)} = R_j^{1/j} \tan^{-1} X_{(j+1)} \qquad (j = n-1, \ldots, 1).$$

7.2 Simulation of stochastic processes

7.2.1 Heterogeneous Poisson process

The homogeneous Poisson process is central to many ideas in stochastic processes. We have already seen (Section 3.3) that to generate times $\{T_{(i)}\}$ of individual events in a Poisson process rate λ, we set $T_{(0)} = 0$, $T_{(i)} = T_{(i-1)} + E_i/\lambda$, where E_i are identically independent negative exponential variates, mean 1.

An important generalization of the Poisson process follows when the rate varies with time. Suppose $N(t_1, t_2)$ denotes the number of events in a time interval (t_1, t_2) and that

$$\text{prob}\{N(t, t+\delta t)>0\} = \lambda(t)\delta t + o(\delta t),$$

$$\text{prob}\{N(t, t+\delta t)>1\} = o(\delta t),$$

where $\lambda(t)$ is a specified non-negative rate function. This describes a hetero-geneous or time-dependent Poisson process. The definition leads to $N(0, t_0)$ being Poisson distributed with integrated rate function and mean $\Delta(t_0) = \int_0^{t_0} \lambda(u)\mathrm{d}u$. From a simulation viewpoint it is usually the timing of the individual events, rather than just the number of events in $[0, t_0]$, which is required. Several examples appear in the literature. Newell (1968) considered queues with time-dependent arrival rates, Lewis (1972) the fluctuation of arrival rates in an intensive care unit, and Gregory (1979) fluctuations in mean call rates in a telephone-answering service in the gas industry. A cumbersome way of simulating such processes is to approximate $\lambda(t)$ by a piecewise uniform function. Such approaches are unnecessary, since exact methods are available. Before discussing these it is noteworthy that poly-nomial representations, $\lambda(t) = \lambda_0 + \lambda_1 t + \ldots + \lambda_n t^n$, are not very useful because the non-negativity of $\lambda(t)$ imposes non-linear constraints on the coefficients $\{\lambda_i\}$. Instead (see, for example, Lewis and Shedler, 1979b) exponential poly-nomial functions, $\lambda(t) = \exp(z(t))$, are preferred, since $\lambda(t)$ never attains negative values. They have the added advantage that simple procedures exist for estimating the coefficients of the polynomial $z(t)$.

Many of the generation methods derive from Lewis and Shedler (1976, 1979a, b). The first, mentioned also by Cox and Miller (1965, p. 154) involves a *time-scale transformation*. Suppose events occur at $\{T_{(i)}\}$ but that a new time scale is defined by $\tau = \int_0^T \lambda(u)\mathrm{d}u$. Then events $\{\tau_{(i)}\}$ in the new time scale follow a homogeneous Poisson process, with unit rate. To generate events from the heterogeneous process in $[0, t_0]$ simulate events $\{\tau_i\}$ from the homogeneous process and solve $\tau_{(i)} = \Delta(T_{(i)})$, subject to $\tau_{(i)} \leqslant \Delta(t_0)$. Thus $T_{(i)} = \Delta^{-1}(\tau_{(i)})$ as long as $\tau_{(i)} \leqslant \Delta(t_0)$. For example, with a log–linear rate function $\lambda(t) = \exp(\alpha + \beta t)$,

$$\Delta(T) = e^{\alpha}(e^{\beta T} - 1)/\beta$$

and

$$T_{(i)} = \beta^{-1} \ln(1 + \beta\tau_{(i)} e^{-\alpha})$$

as long as $\tau_{(i)} \leqslant e^{\alpha}(e^{\beta t_0} - 1)/\beta$. Note that this method requires generation of one exponential variate and one logarithmic evaluation. In general, the usefulness of the method depends upon the ability to invert $\Delta(\cdot)$.

A second method exploits the conditional probability

$$\text{prob}(T_{(i+1)} \leqslant t | T_{(i)} = t_i) = 1 - \exp\{\Delta(t) - \Delta(t_i)\}.$$

Inverting this conditional c.d.f.,

$$\Delta(T_{(i+1)}) - \Delta(T_{(i)}) = -\ln R_{i+1} \qquad (7.6)$$

where $R_{i+1} \sim U(0, 1)$. Equivalently,

$$T_{(i+1)} = \Delta^{-1}(\Delta(T_{(i)}) - \ln R_{i+1}) \qquad (7.7)$$

both expressions subject to $\Delta(T_{(i)}) - \ln R_{i+1} \leqslant \Delta(t_0)$. In comparing this with time-scale transformation, one finds that expression (7.7) is usually more complex to evaluate than $\Delta^{-1}(\tau_{(i)})$. If eqn (7.6) is summed over $i = 0, \ldots, j$, we obtain

$$\Delta(T_{(j+1)}) = -\sum_{i=1}^{j+1} \ln R_i = \tau_{(j+1)},$$

which is identical to time-scale transformation.

A third method is to note that, conditional upon $N(0, t_0) = n$, the times of events in $[0, t_0]$, $\{T_{(i)}\}$, are the n order statistics from the distribution function $\Delta(t)/\Delta(t_0)$, $(0 \leqslant t \leqslant t_0)$. Given the ith uniform order statistic $U_{(i)}$, inversion of the distribution function leads to

$$\frac{\Delta(T_{(i)})}{\Delta(t_0)} = U_{(i)}$$

or

$$T_{(i)} = \Delta^{-1}(\Delta(t_0)U_{(i)}).$$

For example, the log–linear rate function leads to

$$T_{(i)} = \beta^{-1} \ln \{1 + U_{(i)}(e^{\beta t_0} - 1)\}$$

for $i = 1, 2, \ldots, n$. In other cases where $\Delta(\cdot)$ cannot be inverted analytically, sampling from $\Delta(t)/\Delta(t_0)$ may be accomplished using envelope rejection with a target distribution such as

$$g(t) = t_0^{-1} \qquad (0 \leqslant t \leqslant t_0).$$

The efficiency is then governed by

$$M = \max_{0 \leqslant t \leqslant t_0} \left\{ \frac{\lambda(t)}{\Delta(t_0)g(t)} \right\}$$

$$= \frac{\lambda^* t_0}{\Delta(t_0)}$$

where $\lambda^* = \max_{0 \leqslant t \leqslant t_0} \{\lambda(t)\}$. Thus the efficiency can be high if $\lambda(t)$ varies little over $[0, t_0]$. Of course, non-uniform target distributions may yield higher efficiencies. Having obtained n values $T_1, \ldots, T_n$, these must be sorted to give the order statistics $T_{(1)}, \ldots, T_{(n)}$. There are two disadvantages to these order statistic methods. Firstly, a sort of n items is usually required; secondly, a lengthy simulation will make n large and hence impose high memory requirements. One way to avoid these difficulties is to generate the $\{T_{(i)}\}$ sequentially. This should be done in ascending order, otherwise time reversal will necessitate intermediate storage. Using eqn (7.4)

$$1 - \frac{\Delta(T_{(i+1)})}{\Delta(t_0)} = \left\{ 1 - \frac{\Delta(T_{(i)})}{\Delta(t_0)} \right\} R_{i+1}^{1/(n-i)}$$

for $i = 0, 1, \ldots, n-1$, with $\Delta(T_{(0)}) = 0$. This gives

$$T_{(i+1)} = \Delta^{-1}[\Delta(t_0) - \{\Delta(t_0) - \Delta(T_{(i)})\} R_{i+1}^{1/(n-i)}].$$

This convenience of sequential generation is usually achieved at the expense of some computational complexity, as the reader may verify for the log–linear rate function. The preliminary stage of all these order statistic methods is the generation of a Poisson variate, mean $\Delta(t_0)$.

   Thinning methods constitute a fourth approach, and are often useful if the methods above lead to numerical rather than analytical implementations. Suppose we are again interested in generating events in $[0, t_0]$. The thinning method involves the generation of prospective events in a homogeneous Poisson process, rate λ^*, where $\lambda^* \geqslant \lambda(t)$ for all $t \in [0, t_0]$. A prospective event occurring at time t is accepted with probability $\lambda(t)/\lambda^*$. It is a simple matter to show that the 'thinned' process comprising the accepted events is a realization from the heterogeneous process. Algorithm TP shows how successive event times may be printed out. The expected number of prospective events in $[0, t_0]$ is $\lambda^* t_0$, and so the acceptance probability is $\Delta(t_0)/(\lambda^* t_0)$. This is the same as the rejection variant of the order statistic method. In many cases the thinning method is preferred since no sorting or generation of a Poisson variate is required. The efficiency of the thinning process may be

Algorithm TP

Input $[\lambda^*, \lambda(\cdot), t_0]$
Output [   ]

1   $T := 0$.
2   generate $U, V \sim U(0, 1)$.
3   $T := T - \dfrac{\ln U}{\lambda^*}$.
4   If $T > t_0$ exit.
5   If $V \leqslant \lambda(T)/\lambda^*$ then
6   print T
7   End If.
8   Goto 2.

enhanced, of course, by choice of a more adventurous target process with time-dependent rate $\lambda^*(t)$.

The fifth and final approach is due to Lewis and Shedler (1976). They describe a *gap-statistic* method which is peculiar to the log–linear rate function $\lambda(t) = \lambda e^{\alpha_1 t}$, where initially we consider the case $\alpha_1 < 0$. Let M denote a Poisson random variable, mean $-\lambda/\alpha_1$. Then, conditional upon $M = m$, the order statistics $T_{(1)}, \ldots, T_{(m)}$ from an exponential distribution mean $-1/\alpha_1$, form a realization of events from the required heterogeneous process, $\{\lambda(t)\}$. A convenient way to generate the order statistics is to note that the gap-statistics $T_{(i)} - T_{(i-1)}$, (with $T_{(0)} \equiv 0$) are independently distributed as negative exponential variates, mean $1/\{-\alpha_1(m+1-i)\}$, (eqn 7.5). This gives rise to Algorithm GAP for generating events in $[0, t_0]$. Note that the algorithm is suitable only for decreasing rate functions. When $\alpha_1 > 0$, a time-reversal trick can be used to modify GAP. Replace λ by $\lambda e^{\alpha_1 t_0}$ and α_1 by $-\alpha_1$. The ith time printed out now represents the time of the $(M-i+1)$th event during $[0, t_0]$. A disadvantage is that the events are generated in reverse time order, necessitating storage. Since the gap method is quite efficient it is recommended for log–linear rate functions. It can also be used as a basis for generation from non-log–linear rate functions. For example, the rate function could be decomposed into log–linear and residual components. Alternatively, a log–linear rate function could be used as a target process in the thinning method, or as a target distribution in an order statistic implementation, with rejection.

Algorithm GAP

Input $[\alpha_1, \lambda, t_0]$
Output [   ]

1   generate M, a Poisson random variate, mean $-\lambda/\alpha_1$.
2   $T := 0$.
3   For $i := 1, M$
4   generate $R \sim U(0, 1)$
5       $T := T + \ln R/\{\alpha_1(M+1-i)\}$
6   If $T > t_0$ exit.
7   Print T.
8   Next i.
9   Exit.

7.2.2   Two-dimensional Poisson process

This is a natural extension of the Poisson process, where we might, for example, replace events over time by points distributed within a plane. In such a process the number of points in a region of area A is Poisson distributed, with mean λA, independently of the number of events occurring in other non-overlapping regions. λ is the rate of the process. A convenient way to simulate a realization of points is given by Cox and Isham (1980, p. 147). Let $R_{(i)}$ denote the distance of the point which is ith nearest to an arbitrarily chosen origin. Define $A_i = \pi(R_{(i)}^2 - R_{(i-1)}^2)$. This is the area of the annulus between the $(i-1)$th and the ith point. Then

$$\text{prob}(A_i \leqslant x) = 1 - \text{prob}(\text{zero events in a region of area } x)$$

$$= 1 - e^{-\lambda x}.$$

Hence $\pi(R_{(i)}^2 - R_{(i-1)}^2)$ is negative exponentially distributed with mean λ^{-1}. Further the distribution is independent of $R_{(1)}, \ldots, R_{(i-1)}$. If Θ_i is the angular coordinate of the ith point it is intuitively obvious that $\Theta_i \sim U(0, 2\pi)$. Thus one method of generation (in order of increasing distance from the origin) is to deliver

$$R_{(0)} = 0$$

$$R_{(i)} = \left(R_{(i-1)}^2 - \frac{\ln U_i}{\pi \lambda} \right)^{\frac{1}{2}}$$

$$\Theta_i = 2\pi V_i$$

where U_i and V_i are independently $U(0, 1)$. Naturally, if a generated point (R_i, Θ_i) falls outside the region, the point is rejected.

For rectangular and circular regions efficient methods are suggested by Lewis and Shedler (1979a). Consider a two-dimensional process in the plane $\{0 \leqslant X \leqslant x_0, 0 \leqslant Y \leqslant y_0\}$. One approach is first to generate the number of points N, which is Poisson distributed, mean $\lambda x_0 y_0$. Given that $N = n$, the projection of the (conditional) process onto the X- or Y-axis is such that the positions are obtained via order statistics from $U(0, x_0)$ and $U(0, y_0)$, respectively. Thus we may set $X_i = U_i x_0$ and $Y_i = V_i y_0$, where U_i and V_i are independently $U(0, 1)$, for $i = 1, \ldots, n$. Alternatively, the projection of the unconditional process onto the X-axis is a Poisson process, rate λy_0. In this case we deliver $X_{(0)} = 0$, $X_{(i)} = X_{(i-1)} - \{\ln U_i/(\lambda y_0)\}$, $Y_i = y_0 V_i$ until $X_{(i)} > x_0$.

For a circular region of radius r_0, the projection of the process onto the r-axis is a heterogeneous Poisson process with integrated rate function πr^2, $(0 \leqslant r \leqslant r_0)$. This suggests the following method. First generate N, the number of points in the region, as a Poisson variate, mean πr_0^2. Given $N = n$, generate the ordered distances of points from the origin $\{R_{(i)}\}$ as the n order statistics from a c.d.f. $\Delta(r)/\Delta(r_0) = (r/r_0)^2$, $(0 \leqslant r \leqslant r_0)$. An efficient method is to set $R_i = r_0 \max(U_i, V_i)$, where U_i and V_i are independently $U(0, 1)$ for $i = 1, \ldots, n$. Rank the values such that $R_{(1)} \leqslant \ldots \leqslant R_{(n)}$. The angular component, $\Theta_i = 2\pi W_i$ where $W_i \sim U(0, 1)$.

7.2.3   Discrete state Markov processes

We consider here the simulation of discrete state Markov processes where the time parameter may be discrete or continuous. A description of the properties of such processes may be found in any book on stochastic processes, for example Cox and Miller (1965), Bhat (1972).

Let $X(t)$ denote the state of a system at time t, where the state is defined on the non-negative integers. Let t also be defined on the non-negative integers in the discrete time case, and on $[0, \infty)$ in the continuous time case. In the discrete time case suppose the system satisfies a *memoryless* property in the sense that the conditional distribution of $X(t+1)$ given $X(t), X(t-1), \ldots,$ depends only on the value of $X(t)$. Such a system is a *Markov chain* and is specified by transition matrices $P(t) = \{p_{ij}(t)\}$ where the transition probabilities at a step in the process are given by

$$p_{ij}(t) = \text{prob}\{X(t+1)=j \,|\, X(t)=i\}$$

for $i, j = 0, 1, 2, \ldots$. Note that the transition matrices are semi-infinite. This also covers the finite and doubly infinite chains by suitable redefinition of the state space and/or transition probabilities. We consider mainly *homogeneous* Markov chains, that is ones for which $p_{ij}(t) = p_{ij}$ for all t.

Given that a homogeneous Markov chain is in state i, the probability it remains there at the next transition is p_{ii}. Thus the number of transitions, N_i, until a move out of the state is geometrically distributed with

$$\text{prob}(N_i = n) = p_{ii}^{n-1}(1 - p_{ii}) \qquad (n = 1, 2, \ldots).$$

Given a transition that results in a move out of state i, the (conditional) probability that the move is to state j is $p_{ij}/(1 - p_{ii})$. Thus to simulate the process only two discrete variates are needed for each *change* of state.

In the continuous time parameter case, define the transition intensity rates $\{q_{ij}(t)\}$ by

$$\text{prob}(X(t+\delta t)=j \,|\, X(t)=i) = q_{ij}(t)\delta t + o(\delta t) \qquad (i \neq j)$$

and

$$\text{prob}(X(t+\delta t)=i \,|\, X(t)=i) = 1 + q_{ii}(t)\delta t + o(\delta t).$$

In the homogeneous case $q_{ij}(t) = q_{ij}$ for all t. We consider the usual case where $\sum_j q_{ij} = 0$ for all i. [Cox and Miller (1965, p. 180) make reference to processes where $\sum_j q_{ij} \leqslant 0$.] The underlying memoryless property means that given a homogeneous system currently in state i, the time T_i, until it moves out of the state, has p.d.f.

$$f_{T_i}(t) = -q_{ii}e^{q_{ii}t} \qquad (t > 0),$$

while the (conditional) probability that the change is to state j is $-q_{ij}/q_{ii}$.

Algorithm MARKOV provides a simulator for homogeneous processes in discrete and continuous time during $[0, t_0]$. For the continuous time case, p_{ij} is replaced by q_{ij} $(i \neq j)$ and p_{ii} by $1 + q_{ii}$. The initial state $X(0)$ is denoted by k. Note that in the discrete time case the method obviates the need to look at the system at every step. Clearly, if the diagonal elements $\{p_{ii}\}$ are large for those states i, which are visited frequently, then the speed of the simulation is enhanced over that achievable when observing at every step. Steps 11–17 use the 'chop-down search' method mentioned in Section 3.11.1. At the expense of a little extra programming effort, the speed can be improved, for finite Markov chains only of course, by calculating and storing the cumulative probabilities. Step 18 delivers T, the time of the next state change, and i, the new state.

The algorithm can be adapted to cope with heterogeneous processes. In discrete time, the first part of step 3 becomes

$$T = T + 1 \text{ if } R_1 < 1 - p_{ii}(T),$$

Algorithm MARKOV

Input $[t_0, k, \{p_{ij}\}]$
Output [   ]

1    $T := 0$, $i := k$.
2   generate R_1, $R_2 \sim U(0, 1)$.

$$3 \quad T := \begin{cases} \left\langle T + 1 + \dfrac{\ln R_1}{\ln p_{ii}} \right\rangle & \text{(discrete)} \\[2ex] T + \ln R_1 / (p_{ii} - 1) & \text{(continuous)} \end{cases}$$

4   If $T > t_0$ exit.
5    $R_2 := R_2(1 - p_{ii})$.
6   If $i = 0$ then
7       $j := 1$
8   Else
9       $j := 0$.
10   End If.
11   While $R_2 \geqslant p_{ij}$ do
12      $R_2 := R_2 - p_{ij}$
13      $j := j + 1$
14   If $j = i$ then
15         $j := i + 1$
16   End If.
17   End While.
18   $i := j$, Print i, T.
19   Goto 2.

else

$$T = T + v \text{ where } v \text{ is the smallest integer satisfying}$$

$$R_1 < 1 - p_{ii}(T) + \sum_{S=2}^{v} \{1 - p_{ii}(T + S - 1)\} \prod_{j=0}^{S-2} p_{ii}(T + j).$$

Chop-down search can be used to facilitate this. In continuous time, the time till a move out of state i is essentially the time to the next event in a heterogeneous Poisson process, rate $-q_{ii}(t)$. Using eqn (7.7), one possibility is to update the time of the next event by

$$T = \Delta^{-1}\{\Delta(T) - \ln R_1\}$$

where $\Delta(t) = -\int_0^t q_{ii}(u) du$. Depending upon the complexity of $\{-q_{ii}(t)\}$ it may be more appropriate to use 'thinning'.

Returning to the homogeneous case and Algorithm MARKOV, the 'birth–death' processes arise as a special case, where the only state changes possible are $i \to i + 1$ or $i - 1$. So the search for a succeeding state is made faster by initiating the search at $j = \max(i - 1, 0)$ in steps 6–10.

We now give an application of the algorithm. Consider the case of a three-server queueing system. Customers arrive according to a Poisson process rate λ. The three servers have mean service rates μ_1, μ_2, μ_3 where $\mu_1 \geqslant \mu_2 \geqslant \mu_3$. Customer service times are independently and negative exponentially distributed. One model of realistic behaviour is to suppose that customers are 'informed', in which case an arrival goes to the fastest free server. Other decision modes are possible. For example, a customer may not have the benefit of this information and so goes to server i with a specified probability. Rubinovitch (1985) has considered these types of problem for two servers. The interest in the three-server problem considered here is to determine, for fixed λ and $\mu = \mu_1 + \mu_2 + \mu_3$, how the behaviour of the system is dependent on the degree of heterogeneity between servers.

Figure 7.1 shows the state changes which are possible, each arc being labelled with its transition rate intensity. State 0 represents an empty system, states 1, 2, and 3 represent a system with one customer being served by server 1, 2, and 3, respectively. States 4, 5, and 6 indicate that two customers are being served, with servers 3, 2, and 1 being free, respectively. For $j \geqslant 7$, state j represents all servers busy and $j - 4$ customers in the system.

The matrix of transition intensity rates is shown in Fig. 7.2. The matrix is a semi-infinite one. For $i \geqslant 8$, the only non-zero transition probabilities are $p_{ii} = -(\lambda + \mu)$, $p_{i,i-1} = \mu$, $p_{i,i+1} = \lambda$.

Let $X(t)$ denote the state of the system at time t. The system was simulated, using Algorithm MARKOV for $T = 10^5$ time units, with initial condition $X(0) = 0$, mean arrival rate $\lambda = 0.7$, and combined service rate $\mu = \mu_1 + \mu_2 + \mu_3 = 1$. Various combinations of μ_1, μ_2, and μ_3 were tried to investigate the effect of different degrees of heterogeneity. Since the run-

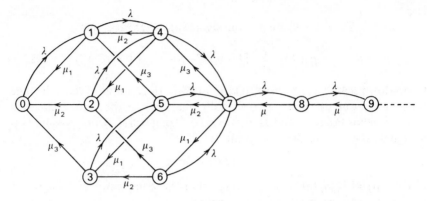

Fig. 7.1. Transition diagram for a three-server (heterogeneous) queueing system

length is extremely long, the transient behaviour near $t=0$ forms only a small part of the total realization. It is therefore reasonable to suppose for the purposes of estimation that $\{X(t): 0 \leqslant t \leqslant T\}$ is approximately a covariance stationary process. If we define $Y(t)$ to be the number of customers in the system at time t, then $\{Y(t): 0 \leqslant t \leqslant T\}$ is also covariance stationary, with $E\{Y(t)\}=\mu_Y$. Accordingly an unbiased estimate of μ_Y is

$$\hat{\mu}_Y = \frac{1}{T} \int_0^T Y(t)\,\mathrm{d}t.$$

Similarly, if p denotes the long-run proportion of time that servers are busy, an unbiased estimate of p is

$$\hat{p} = \frac{1}{3T} \int_0^T \min\{Y(t), 3\}\,\mathrm{d}t.$$

Table 7.1 shows $\hat{\mu}_Y$ and $\hat{p}$ given μ_1, μ_2, and μ_3. For the homogeneous case $\mu_1 = \mu_2 = \mu_3 = 1/3$, analytical considerations give $p=0.7$ and $\mu_Y = 3.25$, which provides a validity check on the simulation model, at least for this special case.

The main feature of interest is that the congestion in the system (as measured by μ_Y and p) increases as the degree of heterogeneity between the servers increases. Notice how the system is easily simulated using Algorithm MARKOV. All that is required is to input the transition intensity rates for $i \leqslant 7$. The alternative is to generate the time of next arrival, and of next departures for each of servers 1, 2, and 3. Some non-trivial coding is required to determine precisely what the state change is at a given event. This involves significantly more programming effort because no use is made of the system's Markov structure.

	0	1	2	3	4	5	6	7	8	9	10	11
0	$-\lambda$	λ	.	.	.	.	.	.	.	.	.	.
1	μ_1	$-(\lambda+\mu_1)$	.	.	λ	.	.	.	.	.	.	.
2	μ_2	.	$-(\lambda+\mu_2)$	.	λ	.	.	.	.	.	.	.
3	μ_3	.	.	$-(\lambda+\mu_3)$	.	λ	.	.	.	.	.	.
4	.	μ_2	μ_1	.	$-(\lambda+\mu_1+\mu_2)$	.	.	λ	.	.	.	.
5	.	μ_3	.	μ_1	.	$-(\lambda+\mu_1+\mu_3)$	.	λ	.	.	.	.
6	.	.	μ_3	μ_2	.	.	$-(\lambda+\mu_2+\mu_3)$	λ	.	.	.	.
7	.	.	.	.	μ_3	μ_2	μ_1	$-(\lambda+\mu)$	λ	.	.	.
8	.	.	.	.	.	.	.	μ	$-(\lambda+\mu)$	λ	.	.
9	.	.	.	.	.	.	.	.	$*$	$*$	$*$	.
10	.	.	.	.	.	.	.	.	.	$*$	$*$	$*$

Fig. 7.2. Transition intensity rate matrix, Q.

Table 7.1 Sample mean number of customers in the system ($\hat{\mu}_Y$) and proportion of time servers are busy ($\hat{p}$) for a $M/M/3$ heterogeneous server queueing system, with $\lambda = 0.7$, and service rates μ_1, μ_2, μ_3

μ_1	μ_2	μ_3	$\hat{\mu}_Y$	$\hat{p}$
0.3$\dot{3}$	0.3$\dot{3}$	0.3$\dot{3}$	3.22	0.70
0.4	0.3	0.3	3.22	0.70
0.4	0.4	0.2	3.27	0.71
0.5	0.25	0.25	3.30	0.71
0.5	0.3	0.2	3.31	0.72
0.5	0.4	0.1	3.43	0.74
0.7	0.15	0.15	3.59	0.77
0.7	0.2	0.1	3.61	0.77
0.9	0.05	0.05	4.03	0.85
0.9	0.09	0.01	4.04	0.86

7.3 Sampling from distribution tails

It is sometimes necessary to sample from the tail of a distribution, such as the normal, gamma, t, Poisson, or negative binomial. Why should tail generation procedures be of use? Firstly, some model building is directed towards understanding the behaviour of a system under stress conditions. In these cases it may be possible to condition the experiment on rare events only, i.e. events occurring in the tail of a distribution. Secondly, procedures for generating from an entire distribution sometimes use a probability mixture (or composition) approach, in which, with a specified probability, variates are generated from the tail. Examples include the rectangle–wedge–tail method for normal generation and the alias algorithm for the Poisson and negative binomial distributions.

7.3.1 Tail generation from the normal distribution

We consider a truncated normal distribution with p.d.f.

$$f_X(x) = e^{-x^2/2} \Big/ \int_a^\infty e^{-u^2/2} \, du \qquad (x \geqslant a),$$

where a is a non-negative constant. Envelope rejection may be used, with a target distribution

$$g_Y(x) = \lambda e^{-\lambda(x-a)} \qquad (x \geqslant a), \tag{7.8}$$

where λ is a parameter whose value is chosen so as to maximize sampling efficiency. We have

$$\frac{f_X(x)}{g_Y(x)} = \frac{e^{-x^2/2} e^{\lambda(x-a)}}{\lambda \int_a^\infty e^{-u^2/2} du} \qquad (x \geqslant a). \qquad (7.9)$$

The optimal value of λ is given by the saddle point of the function $f_X(x)/g_Y(x)$, giving a maximum with respect to x and a minimum with respect to λ. This saddle point occurs at

$$x^* = \lambda^* \qquad (7.10)$$
$$\lambda^* = \{a + \sqrt{(a^2 + 4)}\}/2.$$

Choosing λ in this optimal fashion gives

$$M = \frac{e^{\lambda^{*2}/2 - \lambda^* a}}{\lambda^* \int_a^\infty e^{-u^2/2} du}, \qquad (7.11)$$

which represents the reciprocal sampling efficiency. Prospective variates Y are generated by setting $Y = a + (E_1/\lambda^*)$, where E_1 is a standard negative exponential variate. Given a random number R, (7.9) and (7.10) lead to the acceptance condition

$$R < e^{-\frac{1}{2}(Y - \lambda^*)^2} \qquad (7.12)$$

or

$$E_2 > \tfrac{1}{2}(Y - \lambda^*)^2, \qquad (7.13)$$

where E_2 is a second (independent) standard negative exponential variate. A simple pretest for (7.13) is

$$2(1 - R)/(1 + R) > \tfrac{1}{2}(Y - \lambda^*)^2.$$

If this is not satisfied, then the logarithmic evaluation for E_2 is required. The resulting method is shown in Algorithm TN.

Algorithm TN

Input $[a]$ Saved $[\lambda^*]$
Output $[X]$

S1 $\lambda^* := \{a + \sqrt{(a^2 + 4)}\}/2$.
1   generate $R \sim U(0, 1)$ and E a standard negative exponential variate.
2    $X := a + E/\lambda^*$.
3   If $4(1 - R)/(1 + R) > (X - \lambda^*)^2$ exit.
4   If $-\ln R \leqslant \tfrac{1}{2}(X - \lambda^*)^2$ goto 1.
5   Exit.

At least two other methods are known for sampling from normal tails. If, instead of choosing an optimal value λ^*, a value $\lambda = a$ is used, then the method of Schmeiser (1980) results. In this case the acceptance condition (7.12) becomes

$$R < e^{-\frac{1}{2}(Y-a)^2}, \tag{7.14}$$

where $Y = a - \{(\ln R_1)/a\}$, and $R_1 \sim U(0, 1)$. Since (7.14) can be expressed as

$$R < \frac{e^{-\frac{1}{2}(Y^2 - a^2)}}{e^{-a(Y-a)}},$$

the denominator of the right-hand side being R_1, the acceptance condition becomes $R < \exp\{-0.5(Y^2 - a^2)\}/R_1$. When $Y > a > 1$, a suitable acceptance pretest is $\{1 - a(Y-a)\}/R_1 > R$. Schmeiser's algorithm results in a lower sampling efficiency than that of Algorithm TN (since a non-optimal value of λ is used) and the pretest restricts use of the method to $a \geq 1$.

The other method, due to Marsaglia (1964), is usually known as the tail method. Given a target distribution

$$g_Y(x) = x e^{-\frac{1}{2}(x^2 - a^2)} \qquad (x \geq a), \tag{7.15}$$

$$\frac{f_X(x)}{g_Y(x)} = \frac{e^{-a^2/2}}{x \int_a^\infty e^{-u^2/2} \, du},$$

and

$$M = \frac{e^{-a^2/2}}{a \int_a^\infty e^{-u^2/2} \, du}.$$

Using inversion on (7.15), $Y = \sqrt{(a^2 - 2 \ln R_1)}$, and given a second random number R_2, the acceptance condition becomes $R_2 < a/Y$. Interestingly, the sampling efficiency is identical to Schmeiser's method $[\lambda^* = a$ in eqn (7.11)], but both are uniformly lower than for Algorithm TN, and tend to zero as $a \to 0$. Table 7.2 shows sampling efficiencies (for various truncation points) for the three methods. Computer timings for FORTRAN implementations of the

Table 7.2 Sampling efficiencies for normal tail variate generation

Method	a								
	0	0.01	0.1	0.5	1	2	3	5	10
Optimal exponential	0.760	0.762	0.775	0.828	0.877	0.934	0.961	0.983	0.995
Marsaglia	0	0.012	0.116	0.438	0.656	0.843	0.914	0.964	0.990
Schmeiser	NA	NA	NA	NA	0.656	0.843	0.914	0.964	0.990

NA, not applicable.

Table 7.3 Mean time (μs) to generate a normal tail variate (FORTRAN implementation) on a DEC-20 computer

	a								
Method	0	0.01	0.1	0.5	1	2	3	5	10
Optimal exponential	150	150	148	140	131	120	114	111	108
Marsaglia	—	8555	921	250	172	135	125	119	117
Schmeiser	NA	NA	NA	NA	185	147	136	130	128

NA, not applicable.

three methods are shown in Table 7.3 and reflect closely the sampling efficiencies. We conclude that Algorithm TN provides a fast procedure, which is easy to implement and is robust for all values of a.

7.3.2 Tail generation from the gamma distribution

Dagpunar (1978) shows how to generate from the p.d.f.

$$f_X(x) = \frac{x^{\alpha-1} e^{-x}}{\displaystyle\int_a^\infty v^{\alpha-1} e^{-v} dv} \qquad (x \geqslant a),$$

where $\alpha > 1$, and a is a non-negative constant. The target distribution is negative exponential as in (7.8). This gives

$$\frac{f_X(x)}{g_Y(x)} = \frac{x^{\alpha-1} e^{-\{x(1-\lambda)+\lambda a\}}}{\lambda \displaystyle\int_a^\infty v^{\alpha-1} e^{-v} dv}.$$

The saddle point of this function lies at

$$x^* = (\alpha-1)/(1-\lambda^*),$$
$$\lambda^*(x^*-a) = 1,$$

which gives

$$\lambda^* = [(a-\alpha) + \sqrt{\{(a-\alpha)^2 + 4a\}}]/(2a).$$

With this choice,

$$M = \frac{(\alpha-1)^{\alpha-1} e^{-(\alpha-1+\lambda^*a)}}{(1-\lambda^*)^{\alpha-1} \lambda^* \displaystyle\int_a^\infty v^{\alpha-1} e^{-v} dv}.$$

The condition for acceptance becomes

$$R_2 < \left\{ \frac{Y(1-\lambda^*)}{\alpha-1} \right\}^{\alpha-1} e^{\alpha-1-Y(1-\lambda^*)}$$

or

$$-\ln R_2 > (1-\lambda^*)\,Y - (\alpha-1)\left\{ 1 + \ln Y + \ln\left(\frac{1-\lambda^*}{\alpha-1}\right) \right\},$$

which results in Algorithm TG.

Algorithm TG

Input $[a, \alpha]$ Saved $[\lambda^*, b, c]$
Output $[X]$

S1 $\lambda^* := [(a-\alpha) + \sqrt{\{(a-\alpha)^2 + 4a\}}]/(2a)$.
S2 $b := 1 - \lambda^*$
S3 $c := \ln[b/(\alpha-1)]$.
1   generate $R_1, R_2 \sim U(0, 1)$.
2     $X := a - (\ln R_1)/\lambda^*$.
3     $S := bX - (\alpha-1)(1 + \ln X + c)$
4   If $2(1 - R_2)/(1 + R_2) > S$ exit.
5   If $-\ln R_2 \leqslant S$ goto 1.
6   Exit.

For fixed parameter values λ^*, b, and c are obviously saved between calls. Step 4 is an acceptance pretest. Table 7.4 gives timings for various shape parameter values, α, and truncation points, t_p, where t_p is the tail probability. The speed is fairly robust with respect to α and t_p, showing that in most cases the optimal exponential provides a good fit. Where the truncation point is

Table 7.4 Mean time (μs) to generate a gamma tail variate (FORTRAN implementation) on a DEC-20 computer

	α			
t_p	$1 + \varepsilon^\dagger$	3	10	50
$1 - \varepsilon^\dagger$	157	310	621	1445
0.75	160	195	228	244
0.50	154	181	196	205
0.25	154	168	173	180
0.10	157	165	168	174
0.05	161	163	168	174
0.01	163	160	162	166

$^\dagger \varepsilon = 0+$

very much smaller than the mode, the routine becomes slower, due to a poorer fit. Fortunately, in such circumstances a method based on sampling from the entire distribution, with rejection of values lower than the truncation point, gives good efficiency.

7.3.3   Tail generation from a t-distribution

Unfortunately, it is not possible to extend the methods used previously to the t-distribution. This is because, apart from the asymptotic normal case, no exponential target distribution can envelope a t-distribution. One possibility is to use a truncated Cauchy as a target distribution. For $n=1$, this would clearly lead to a 100% sampling efficiency. For other values of n, however, determination of optimal parameter values for such a distribution requires numerical methods, which do not lead to a convenient implementation.

However, the ratio of uniforms approach is eminently suitable for tail variate generation, (see Dagpunar, 1988). A t-distribution, with n degrees of freedom, truncated at a ($\geqslant 0$) has p.d.f.

$$f_X(x)=k(a)\left(1+\frac{x^2}{n}\right)^{-(n+1)/2} \qquad (x\geqslant a)$$

where

$$k(a)=\{\sqrt{(n)}B(\tfrac{1}{2},\, n/2)P(X_n\geqslant a)\}^{-1}$$

and X_n denotes an untruncated t-variate. Relocating the distribution, set $Y=X-a$, giving

$$f_Y(y)=k(a)\left\{1+\frac{(y+a)^2}{n}\right\}^{-(n+1)/2} \qquad (y\geqslant 0).$$

The reciprocal variate $Z=Y^{-1}$ has density

$$f_Z(z)=k(a)z^{-2}\left\{1+\frac{(z^{-1}+a)^2}{n}\right\}^{-(n+1)/2} \qquad (z\geqslant 0).$$

The modes of these two distributions lie at

$$m_Y=0$$

and

$$m_Z=\begin{cases} 0 & (n=1) \\ \dfrac{(n-3)a+\sqrt{\{(n-3)^2a^2+8(n-1)(a^2+n)\}}}{4(a^2+n)} & (n>1) \end{cases}$$

Thus the region, $C=\{(u,v): 0\leqslant u\leqslant f_Y^{\frac{1}{2}}(v/u)\}$ is totally enclosed by

$D = \{(u, v): 0 \leqslant u \leqslant u_+; 0 \leqslant v \leqslant v_+\}$ where

$$u_+ = f_Y^{\frac{1}{2}}(m_Y) = k^{\frac{1}{2}}(a)\left(1 + \frac{a^2}{n}\right)^{-(n+1)/4}$$

and

$$v_+ = f_Z^{\frac{1}{2}}(m_Z) = \begin{cases} k^{\frac{1}{2}}(a) & (n = 1) \\ k^{\frac{1}{2}}(a)\{1 + (m_Z^{-1} + a)^2/n\}^{-(n+1)/4}/m_Z & (n > 1). \end{cases}$$

Given two random numbers R_1 and R_2, $Y = R_2 v_+ /(R_1 u_+)$ is delivered subject to $R_1 u_+ \leqslant f_Y^{\frac{1}{2}}(Y)$. Transforming back to the truncated variate, X, results in algorithm TT. The sampling efficiency of the algorithm is $P = 0.5/(u_+ v_+)$. Specimen values are shown in Table 7.5. The first row corresponds to a folded t-distribution, where the efficiencies are identical to those of Algorithm T1, considered in Section 4.5.2. The lowest sampling efficiency is 0.5, for the case $n = 1$, $t_p \to 0$. Typical computer timings for a FORTRAN implementation on a DEC-20 were 228 μs ($n = 4$, $t_p = 0.1$) and 234 μs ($n = 60$, $t_p = 0.001$).

Table 7.5 Sampling efficiencies for tail variate generation from a t-distribution with n degrees of freedom and with tail probability t_p

	n			
t_p	1	4	10	∞
0.5	0.785	0.773	0.748	0.731
0.1	0.508	0.695	0.695	0.690
0.01	0.500	0.678	0.685	0.682
0.001	0.500	0.674	0.682	0.681

Algorithm TT

Input $[n, a]$　　Saved $[g]$
Output $[X]$

S1　If $n \leqslant 1$ then
S2　　$g := \sqrt{(1 + a^2)}$
S3　Else
S4　　$m_Z := \dfrac{(n-3)a + \sqrt{\{(n-3)^2 a^2 + 8(n-1)(a^2 + n)\}}}{4(a^2 + n)}.$

S5　　$g := \{(n + a^2)/(n + [m_Z^{-1} + a]^2)\}^{(n+1)/4}/m_Z.$
S6　End If.
1　generate $R_1, R_2 \sim U(0, 1)$.
2　$X := gR_2/R_1 + a$.
3　If $-\ln R_1 \leqslant \left(\dfrac{n+1}{4}\right)\ln\left(\dfrac{n + X^2}{n + a^2}\right)$ goto 1.

4　Exit.

Appendix 1
FORTRAN 77 routines for generating variates from selected distributions

1.1 RND( ), random normal deviate

```
00100          FUNCTION RND( )
00200    C
00300    C     FUNCTION GENERATES A STANDARD RANDOM NORMAL
00400    C     DEVIATE, USING THE POLAR BOX MULLER METHOD.
00500    C
00600          LOGICAL I
00700          SAVE I,B,U2
00800          DATA I/.FALSE./
00900          IF (I) THEN
01000             RND=B*U2
01100          ELSE
01200    10       U1=2.0*RAN( )-1.0
01300             U2=2.0*RAN( )-1.0
01400             S=U1*U1+U2*U2
01500             IF(S.GE.1.0.OR.S.LE.0.0) GOTO 10
01600             B=SQRT(-2.0*ALOG(S)/S)
01700             RND=B*U1
01800          END IF
01900          I=.NOT.I
02000          RETURN
02100          END
```

1.2 GAMMA(S), gamma calling routine

```
00100           FUNCTION GAMMA(S)
00200    C
00300    C FUNCTION GENERATES A RANDOM GAMMA VARIATE,
00400    C CALLING EITHER GAMMA1 (S>1.0) OR
00500    C EXPON (S=1.0) OR GAMMA2 (S<1.0).
00600    C
00700    C   S=SHAPE PARAMETER OF DISTRIBUTION (0<REAL).
00800    C
00900          IF(S.LE.0.0)THEN
01000             PRINT*,'SHAPE PARAMETER VALUE MUST
01100          *BE POSITIVE'
01200             STOP
01300          END IF
01400          IF(S.GT.1.0)THEN
01500             GAMMA=GAMMA1(S)
01600          ELSE IF(S.LT.1.0)THEN
01700             GAMMA=GAMMA2(S)
01800          ELSE
01900             GAMMA=EXPON(S)
02000          END IF
02100          RETURN
02200          END
```

1.3 GAMMA1(S), gamma (S>1)

```
00100          FUNCTION GAMMA1(S)
00200    C
00300    C FUNCTION GENERATES A RANDOM VARIATE IN
00400    C [0,INFINITY) FROM A GAMMA DISTRIBUTION
00500    C WITH DENSITY PROPORTIONAL TO
00600    C GAMMA1**(S-1)*EXP(-GAMMA1),
00700    C USING BEST'S T DISTRIBUTION METHOD
00800    C
00900    C    S-SHAPE PARAMETER OF DISTRIBUTION
01000    C         (1.0<REAL)
01100    C
01200    C
01300          SAVE B,H,SS
01400          DATA SS/0.0/
01500          IF(S.LE.1.0)THEN
01600              PRINT*,'IMPERMISSIBLE SHAPE PARAMETER VALUE'
01700              STOP
01800          END IF
01900          IF(S.NE.SS)THEN
02000              B=S-1.0
02100              H=SQRT(3.0*S-0.75)
02200              SS=S
02300          END IF
02400    10    R=RAN( )
02500          G=R-R*R
02600          IF(G.LE.0.0)GOTO 10
02700          F=(R-0.5)*H/SQRT(G)
02800          X=B+F
02900          IF(X.LE.0.0)GOTO 10
03000          D=64.0*G*(RAN( )*G)**2
03100          IF(D.LE.0.0)GOTO 20
03200          IF(D*X.LT.X-2.0*F*F)GOTO 20
03300          IF(ALOG(D).GE.2.0*(B*ALOG(X/B)-F))GOTO 10
03400    20    GAMMA1=X
03500          RETURN
03600          END
```

1.4 GAMMA2(S), gamma ($S<1$)

```
00100          FUNCTION GAMMA2(S)
00200    C
00300    C FUNCTION GENERATES A RANDOM VARIATE IN
00400    C [0,INFINITY) FROM A GAMMA DISTRIBUTION
00500    C WITH DENSITY PROPORTIONAL TO
00600    C GAMMA2**(S-1)*EXP(-GAMMA2),
00700    C USING A SWITCHING METHOD.
00800    C
00900    C    S=SHAPE PARAMETER OF DISTRIBUTION
01000    C          (REAL<1.0)
01100    C
01200    C THE PARAMETER U IS THE MACHINE'S SMALLEST
01300    C REAL NUMBER.
01400    C
01500          PARAMETER(U=1E-37)
01600          SAVE A,P,C,UF,SS,VR,D
01700          DATA SS/10.0/
01800          IF(S.LE.0.0.OR.S.GE.1.0)THEN
01900               PRINT*,'SHAPE PARAMETER VALUE OUTSIDE
02000     *PERMITTED RANGE'
02100               STOP
02200          END IF
02300          IF(S.NE.SS)THEN
02400               A=1.0-S
02500               P=A/(A+S*EXP(-A))
02600               IF(P.EQ.1.0)THEN
02700                    PRINT*,'SHAPE PARAMETER VALUE TOO SMALL'
02800                    STOP
02900               END IF
03000               C=1.0/S
03100               UF=P*(U/A)**S
03200               VR=1.0-U
03300               D=A*ALOG(A)
03400               SS=S
03500          END IF
03600    2     R=RAN( )
03700          IF(R.GE.VR)THEN
03800               GOTO 2
03900          ELSE IF(R.GT.P)THEN
04000               X=A-ALOG((1.0-R)/(1.0-P))
04100               W=A*ALOG(X)-D
04200          ELSE IF(R.GT.UF)THEN
04300               X=A*(R/P)**C
04400               W=X
04500          ELSE
04600               GAMMA2=0.0
04700               RETURN
04800          END IF
04900          R=RAN( )
05000          IF(1.0-R.LE.W.AND.R.GT.0.0)THEN
05100               IF(R*(W+1.0).GE.1.0)GOTO 2
05200               IF(-ALOG(R).LE.W)GOTO 2
05300          END IF
05400          GAMMA2=X
05500          RETURN
05600          END
```

1.5   EXPON( ), exponential

```
00100          FUNCTION EXPON( )
00200     C
00300     C FUNCTION GENERATES A RANDOM VARIATE IN
00400     C [0,INFINITY) FROM A NEGATIVE EXPONENTIAL
00500     C DISTRIBUTION WITH DENSITY PROPORTIONAL
00600     C TO EXP(-EXPON),USING INVERSION.
00700     C
00800     1    R=RAN( )
00900          IF(R.LE.0.0)GOTO 1
01000          EXPON=-ALOG(R)
01100          RETURN
01200          END
```

1.6   BETA(AA, BB), beta

```
00100          FUNCTION BETA(AA,BB)
00200     C
00300     C FUNCTION GENERATES A RANDOM VARIATE IN [0,1]
00400     C FROM A BETA DISTRIBUTION WITH DENSITY
00500     C PROPORTIONAL TO BETA**(AA-1)*(1-BETA)**(BB-1),
00600     C USING CHENG'S LOG LOGISTIC METHOD.
00700     C
00800     C    AA=SHAPE PARAMETER FROM DISTRIBUTION (0<REAL)
00900     C    BB=SHAPE PARAMETER FROM DISTRIBUTION (0<REAL)
01000     C
01100     C THE PARAMETERS U AND V REPRESENT THE MACHINE'S
01200     C SMALLEST AND LARGEST REAL NUMBERS RESPECTIVELY.
01300     C
01400          PARAMETER(U=1E-37,V=1E38,ALN4=1.3862944)
01500          LOGICAL SWAP
01600          SAVE D,F,H,T,C,A,B,SWAP
01700          DATA A,B/-1.0,-1.0/
01800          IF(AA.LE.0.0.OR.BB.LE.0.0)THEN
01900               PRINT*,'IMPERMISSIBLE SHAPE PARAMETER VALUE(S)'
02000               STOP
02100          END IF
02200          IF(AA.NE.A.OR.BB.NE.B)THEN
02300               A=AA
02400               B=BB
02500               SWAP=B.GT.A
02600               IF(SWAP)THEN
02700                    G=B
02800                    B=A
02900                    A=G
03000               END IF
03100               D=A/B
03200               F=A+B
03300               IF(B.GT.1.0)THEN
03400                    H=SQRT((2.0*A*B-F)/(F-2.0))
03500                    T=1.0
03600               ELSE
03700                    H=B
03800                    T=1.0/(1.0+(A/(V*B))**B)
03900               END IF
04000               C=A+H
04100               A=AA
04200               B=BB
04300          END IF
```

```
04400    7    R=RAN( )
04500         S=R*R*RAN( )
04600         IF(R.LT.U.OR.S.LE.0.0)GOTO 7
04700         IF(R.LT.T)THEN
04800              X=ALOG(R/(1.0-R))/H
04900              Y=D*EXP(X)
05000              Z=C*X+F*ALOG((1.0+D)/(1.0+Y))-ALN4
05100              IF(S-1.0.GT.Z)THEN
05200                   IF(S-S*Z.GT.1.0)GOTO 7
05300                   IF(ALOG(S).GT.Z)GOTO 7
05400              END IF
05500              BETA=Y/(1.0+Y)
05600         ELSE
05700              IF(4.0*S.GT.(1.0+1.0/D)**F)GOTO 7
05800              BETA=1.0
05900         END IF
06000         IF(SWAP)BETA=1.0-BETA
06100         RETURN
06200         END
```

1.7   TDISTN(M), Student's *t*

```
00100         FUNCTION TDISTN(M)
00200    C
00300    C FUNCTION GENERATES A RANDOM VARIATE FROM A
00400    C T DISTRIBUTION,USING KINDERMAN AND MONAHAN'S
00500    C RATIO METHOD.
00600    C
00700    C    M=DEGREES OF FREEDOM OF DISTRIBUTION
00800    C         (1<=INTEGER)
00900    C
01000         SAVE S,C,A,F,G,MM
01100         DATA MM/0/
01200         IF(M.LT.1)THEN
01300              PRINT*,'IMPERMISSIBLE DEGREES OF FREEDOM'
01400              STOP
01500         END IF
01600         IF(M.NE.MM)THEN
01700              S=M
01800              C=-0.25*(S+1.0)
01900              A=4.0/(1.0+1.0/S)**C
02000              F=16.0/A
02100              IF(M.GT.1)THEN
02200                   G=S-1.0
02300                   G=((S+1.0)/G)**C*SQRT((S+S)/G)
02400              ELSE
02500                   G=1.0
02600              END IF
02700              MM=M
02800         END IF
02900    2    R=RAN( )
03000         IF(R.LE.0.0)GOTO 2
03100         X=(2.0*RAN( )-1.0)*G/R
03200         V=X*X
03300         IF(V.GT.5.0-A*R)THEN
03400              IF(M.GE.3.AND.R*(V+3.0).GT.F)GOTO 2
03500              IF(R.GT.(1.0+V/S)**C)GOTO 2
03600         END IF
03700         TDISTN=X
03800         RETURN
03900         END
```

1.8   GIG (H, B), generalized inverse Gaussian

```
00100          FUNCTION GIG(H,B)
00200   C
00300   C FUNCTION GENERATES A RANDOM VARIATE IN
00400   C [0,INFINITY] FROM A REPARAMETERISED
00500   C GENERALISED INVERSE GAUSSIAN DISTRIBUTION
00600   C WITH DENSITY PROPORTIONAL TO
00700   C GIG**(H-1)*EXP(-0.5*B*(GIG+1/GIG))
00800   C USING A RATIO METHOD.
00900   C
01000   C    H=PARAMETER OF DISTRIBUTION (0<=REAL)
01100   C    B=PARAMETER OF DISTRIBUTION (0<REAL)
01200   C
01300   C THE PARAMETER V IS THE MACHINE'S LARGEST
01400   C REAL NUMBER.
01500   C
01600          PARAMETER(V=1E38)
01700          SAVE A,C,D,E,HH,BB
01800          DATA HH,BB/-1.0,-1.0/
01900          IF(H.LT.0.0.OR.B.LE.0.0)THEN
02000              PRINT*,'IMPERMISSIBLE DISTRIBUTION
02100     *PARAMETER VALUES'
02200              STOP
02300          END IF
02400          IF(H.NE.HH.OR.B.NE.BB)THEN
02500              IF(H.GT.0.25*B*SQRT(V))THEN
02600                  PRINT*,'THE RATIO H:B IS TOO LARGE'
02700                  STOP
02800              END IF
02900              E=B*B
03000              D=H+1.0
03100              YM=(-D+SQRT(D*D+E))/B
03200              IF(YM.EQ.0.0)THEN
03300                  PRINT*,'THE VALUE OF B IS TOO SMALL'
03400                  STOP
03500              END IF
03600              D=H-1.0
03700              XM=(D+SQRT(D*D+E))/B
03800              D=0.5*D
03900              E=-0.25*B
04000              R=XM+1.0/XM
04100              W=XM*YM
04200              A=W**(-0.5*H)*SQRT(XM/YM)*EXP(-E*(R-YM-1.0/YM))
04300              IF(A.EQ.0.0)THEN
04400                  PRINT*,'THE VALUE OF H IS TOO LARGE'
04500                  STOP
04600              END IF
04700              C=-D*ALOG(XM)-E*R
04800              HH=H
04900              BB=B
05000          END IF
05100   2      R1=RAN( )
05200          IF(R1.LE.0.0)GOTO 2
05300          R2=RAN( )
05400          X=A*R2/R1
05500          IF(X.LE.0.0)GOTO 2
05600          IF(ALOG(R1).GT.D*ALOG(X)+E*(X+1.0/X)+C)GOTO 2
05700          GIG=X
05800          RETURN
05900          END
```

1.9 VMISES(SK), von Mises

```
00100          FUNCTION VMISES(SK)
00200   C
00300   C FUNCTION GENERATES A RANDOM VARIATE IN THE RANGE
00400   C [-PI,PI] FROM A VON MISES DISTRIBUTION WITH
00500   C DENSITY PROPORTIONAL TO
00600   C EXP(SK*COS(VMISES))
00700   C USING BEST AND FISHER'S METHOD.
00800   C
00900   C    SK=PARAMETER OF DISTRIBUTION (0<REAL)
01000   C
01100   C THE PARAMETER V IS THE MACHINE'S LARGEST REAL NUMBER.
01200   C
01300          SAVE R,SS
01400          PARAMETER(PI=3.141592654,V=1E38)
01500          DATA SS/-1.0/
01600          IF(SK.LE.0.0)THEN
01700               PRINT*,'IMPERMISSIBLE DISTRIBUTION PARAMETER VALUES'
01800               STOP
01900          END IF
02000          IF(SK.NE.SS)THEN
02100               R=1.0+SQRT(1.0+4.0*SK*SK)
02200               R=(R-SQRT(R+R))/(SK+SK)
02300               IF(R.EQ.0.0)THEN
02400                    R=1.0/SK
02500               ELSE
02600                    R=(1.0+R*R)/(R+R)
02700               END IF
02800               SS=SK
02900          END IF
03000   24     F=COS(PI*RAN( ))
03100          F=(1.0+R*F)/(R+F)
03200          IF(F.GT.1.0)THEN
03300               F=1.0
03400          ELSE IF(F.LT.-1.0)THEN
03500               F=-1.0
03600          END IF
03700          C=SK*(R-F)
03800          IF(C.LE.0.0)GOTO 24
03900          W=RAN( )
04000          IF(C*(2.0-C).LE.W)THEN
04100               IF(C.LE.V*W)THEN
04200                    IF(ALOG(C/W)+1.0.LT.C)GOTO 24
04300               ELSE IF(W.GT.0.0)THEN
04400                    IF(ALOG(C)-ALOG(W)+1.0.LT.C)GOTO 24
04500               END IF
04600          END IF
04700          VMISES=ACOS(F)
04800          IF(RAN( ).LT.0.5)VMISES=-VMISES
04900          RETURN
05000          END
```

1.10 IPOIS(E), Poisson calling routine

```
00100        FUNCTION IPOIS(E)
00200     C
00300     C FUNCTION GENERATES A RANDOM POISSON VARIATE
00400     C CALLING EITHER IPOIS1(E) (E<EC) OR IPOIS2(E)
00500     C (E>=EC).
00600     C
00700     C    E=MEAN OF POISSON DISTRIBUTION (0<REAL)
00800     C
00900     C THE PARAMETER EC MAY BE SET TO ANY VALUE
01000     C EXCEEDING 1. TO ENSURE THAT THE FASTER OF
01100     C IPOIS1(E),IPOIS2(E) IS CALLED EC MAY BE
01200     C DETERMINED BY EXPERIMENTATION. OTHERWISE
01300     C A VALUE OF 20.0 IS SUITABLE.
01400     C
01500        PARAMETER(EC=20.0)
01600        IF(E.LE.0.0)THEN
01700            PRINT*,'IMPERMISSIBLE DISTRIBUTION PARAMETER VALUE'
01800            STOP
01900        END IF
02000        IF(E.LT.EC)THEN
02100            IPOIS=IPOIS1(E)
02200        ELSE
02300            IPOIS=IPOIS2(E)
02400        END IF
02500        RETURN
02600        END
```

1.11   IPOIS1(E), Poisson (*E* < 20)

```
00100          FUNCTION IPOIS1(E)
00200   C
00300   C FUNCTION GENERATES A RANDOM POISSON VARIATE,
00400   C USING UNSTORED INVERSION.
00500   C
00600   C    E=MEAN OF POISSON DISTRIBUTION (0<REAL)
00700   C
00800   C THE PARAMETER ULN=-ALOG(MACHINE'S
00900   C SMALLEST REAL NUMBER)
01000   C
01100          PARAMETER(ULN=85.19)
01200          SAVE W,EE
01300          DATA EE/-1.0/
01400          IF(E.GT.ULN)THEN
01500              PRINT*,'E IS TOO LARGE FOR UNSTORED INVERSION'
01600              STOP
01700          END IF
01800          IF(E.NE.EE)THEN
01900              W=EXP(-E)
02000              EE=E
02100          END IF
02200          X=0.0
02300          Y=W
02400          R=RAN( )
02500   5      IF(R.GT.Y)THEN
02600              X=X+1.0
02700              R=R-Y
02800              Y=E*Y/X
02900              GOTO 5
03000          END IF
03100          IPOIS1=X+0.5
03200          RETURN
03300          END
```

1.12   IPOIS2(E), Poisson (*E* ⩾ 20)

```
00100          FUNCTION IPOIS2(E)
00200   C FUNCTION GENERATES A RANDOM POISSON VARIATE,
00300   C USING A MODIFIED VERSION OF ATKINSON'S
00400   C LOGISTIC ENVELOPE METHOD.
00500   C
00600   C    E=MEAN OF POISSON DISTRIBUTION (1<=REAL)
00700   C
00800   C THE PARAMETER U IS THE MACHINE'S SMALLEST
00900   C REAL NUMBER.
01000   C THE PARAMETER VF IS THE MACHINE'S LARGEST
01100   C REAL NUMBER.
01200   C THE PARAMETER EMAX=3*(ALOG(VF)/PI)**2.
01300   C
01400   C IF J IS THE SMALLEST POSITIVE INTEGER FOR
01500   C WHICH HB(J) EXCEEDS E THEN THE RECIPROCAL
01600   C OF HA(J) REPRESENTS THE ACCEPTANCE
01700   C PROBABILITY MULTIPLIED BY SQRT(1.5/PI**3).
01800   C
```

```
01900          SAVE RBET,GA,DEL,F,EE,RC
02000          PARAMETER(PI=3.1415927,PI2=6.2831853,U=1E-37,EMAX=2327.12)
02100          PARAMETER(VF=1E38)
02200          DIMENSION HA(10),HB(10)
02300          DATA HA/22.069,13.249,9.316,7.729,7.029,
02400     *         6.747,6.461,6.183,6.015,5.801/
02500          DATA HB/2.0,5.0,10.0,20.0,30.0,
02600     *         50.0,100.0,200.0,1000.0,1E38/
02700          DATA EE/-1.0/
02800          IF(E.NE.EE)THEN
02900               IF(E.LT.1.0)THEN
03000                    PRINT*,'E IS TOO SMALL FOR LOGISTIC
03100     *ENVELOPE METHOD'
03200                    STOP
03300               END IF
03400               IF(E.GT.EMAX)THEN
03500                    PRINT*,'E IS TOO LARGE'
03600                    STOP
03700               END IF
03800               DO 12 J=1,10
03900               IF(HB(J).GT.E)GOTO 14
04000     12        CONTINUE
04100               J=10
04200     14        RBET=SQRT(3.0*E)/PI
04300               GA=EXP((0.5+E)/RBET)
04400               RC=1.0/(1.0+GA/VF)
04500               DEL=-E+ALOG((1.0+GA)/HA(J))
04600               F=0.5*ALOG(E*PI2)
04700               EE=E
04800          END IF
04900     15   R=RAN( )
05000          V=0.0
05100          Z=1.0+GA*R
05200          Y=1.0-R
05300          IF(Y.LE.U)GOTO 15
05400          IF(R.LT.RC)THEN
05500               WP=RBET*ALOG(Z/Y)
05600          ELSE
05700               WP=E+0.5+ALOG(R/(1.0-R))*RBET
05800          END IF
05900          IX=WP
06000          W=RAN( )*Y*Z
06100          IF(W.LE.0.0)GOTO 15
06200          W=ALOG(W)
06300          IF(IX.GT.0)THEN
06400               Y=IX
06500               Z=W+(Y+0.5)*ALOG(Y/E)-Y
06600               IF(Z+1.0/(12.0*Y).LE.DEL)GOTO 30
06700               IF(Z.GT.DEL)GOTO 15
06800               Y=1.0/E
06900               Z=Y
07000               DO 20 J=1,IX
07100               V=V+ALOG(Z)
07200               Z=Z+Y
07300     20        CONTINUE
07400          END IF
07500          IF(W+V.GT.DEL+F)GOTO 15
07600     30   IPOIS2=IX
07700          RETURN
07800          END
```

1.13 IBINOM(PS, N), binomial

```
00100          FUNCTION IBINOM(PS,N)
00200    C
00300    C FUNCTION GENERATES A RANDOM BINOMIAL
00400    C VARIATE USING UNSTORED INVERSION.
00500    C
00600    C    PS=BERNOULLI SUCCESS PROBABILITY
00700    C           (0<=REAL<=1)
00800    C    N=NUMBER OF BERNOULLI TRIALS
00900    C           (1<=INTEGER)
01000    C
01100    C THE PARAMETER US=-ALOG10(MACHINE'S
01200    C SMALLEST REAL NUMBER).
01300    C
01400          PARAMETER(US=37.0)
01500          LOGICAL B
01600          SAVE P,W
01700          DATA PT,NT/-1.0,0.0/
01800          IF(N.LT.1.OR.PS.LT.0.0.OR.PS.GT.1.0)THEN
01900              PRINT*,'IMPERMISSIBLE DISTRIBUTION PARAMETER VALUES'
02000              STOP
02100          END IF
02200          IF(PT.NE.PS.OR.NT.NE.N)THEN
02300              B=PS.LE.0.5
02400              IF(B)THEN
02500                  P=PS
02600                  Q=1.0-PS
02700              ELSE
02800                  P=1.0-PS
02900                  Q=PS
03000              END IF
03100              IF(N*P.GT.1.66096*US)THEN
03200                  PRINT*,'DISTRIBUTION MEAN IS EITHER TOO
03300      *LARGE OR TOO SMALL'
03400                  STOP
03500              END IF
03600              W=Q**N
03700              P=P/Q
03800              PT=PS
03900              NT=N
04000          END IF
04100          R=RAN( )
04200          X=0.0
04300          Y=W
04400          G=N
04500    5     IF(Y.LT.R)THEN
04600              R=R-Y
04700              X=X+1.0
04800              Y=Y*P*G/X
04900              G=G-1.0
05000              GOTO 5
05100          END IF
05200          IF(B)THEN
05300              IBINOM=X+0.5
05400          ELSE
05500              IBINOM=N-X+0.5
05600          END IF
05700          RETURN
05800          END
```

1.14   NBINOM(SK, P), negative binomial

```
00100            FUNCTION NBINOM(SK,P)
00200      C FUNCTION GENERATES A RANDOM NEGATIVE
00300      C BINOMIAL VARIATE USING UNSTORED
00400      C INVERSION AND OR THE REPRODUCTIVE
00500      C PROPERTY.
00600      C
00700      C   SK='NUMBER' OF FAILURES REQUIRED
00800      C            (0<REAL)
00900      C   P=BERNOULLI SUCCESS PROBABILITY
01000      C            (0<REAL<1)
01100      C
01200      C THE PARAMETER H IS SET SO THAT
01300      C UNSTORED INVERSION ONLY IS USED
01400      C WHEN P<=H, OTHERWISE A COMBINATION OF
01500      C UNSTORED INVERSION AND THE REPRODUCTIVE
01600      C PROPERTY IS USED.
01700      C THE PARAMETER ULN=-ALOG(MACHINE'S SMALLEST REAL NUMBER)
01800      C
01900            PARAMETER(H=0.7,ULN=85.19)
02000            IF (SK.LE.0.0.OR.P.LE.0.0.OR.P.GE.1.0)THEN
02100               PRINT*,'IMPERMISSIBLE DISTRIBUTION PARAMETER VALUES'
02200               STOP
02300            END IF
02400            Q=1.0-P
02500            X=0.0
02600            ST=SK
02700            IF(P.GT.H)THEN
02800               V=1.0/ALOG(P)
02900               K=ST
03000               DO 20 I=1,K
03100      10        R=RAN( )
03200               IF(R.LE.0.0)GOTO 10
03300               N=V*ALOG(R)
03400               X=X+N
03500      20        CONTINUE
03600               ST=ST-K
03700            END IF
03800            S=0.0
03900            IF(ST.GT.-ULN/ALOG(Q))THEN
04000               PRINT*,' P IS TOO LARGE FOR THIS VALUE OF SK'
04100               STOP
04200            END IF
04300            Y=Q**ST
04400            G=ST
04500            R=RAN( )
04600      30   IF(Y.LT.R)THEN
04700               R=R-Y
04800               S=S+1.0
04900               Y=Y*P*G/S
05000               G=G+1.0
05100               GOTO 30
05200            END IF
05300            NBINOM=X+S+0.5
05400            RETURN
05500            END
```

1.15 MVNORM(N, H, D, F, FIRST, X), multivariate normal

```
00100          SUBROUTINE MVNORM(N,H,D,F,FIRST,X)
00200    C
00300    C SUROUTINE GENERATES AN N VARIATE RANDOM NORMAL
00400    C VECTOR USING A CHOLESKY DECOMPOSITION, CALLING
00500    C FUNCTION RND( ) WHICH RETURNS A STANDARD RANDOM
00600    C NORMAL DEVIATE.
00700    C
00800    C ARGUMENTS:
00900    C          N=NUMBER OF VARIATES IN VECTOR
01000    C              (INPUT,INTEGER>=1)
01100    C        H(J)=J'TH ELEMENT OF VECTOR OF MEANS
01200    C              (INPUT,REAL)
01300    C        X(J)=J'TH ELEMENT OF DELIVERED VECTOR
01400    C              (OUTPUT,REAL)
01500    C
01600    C      D(J*(J-1)/2+I)=(I,J)'TH ELEMENT OF VARIANCE
01700    C              MATRIX (J>=I)
01800    C                (INPUT,REAL)
01900    C      F((J-1)*(2*N-J)/2+I)=(I,J)'TH ELEMENT OF LOWER TRIANGULAR
02000    C              DECOMPOSITION OF VARIANCE
02100    C              MATRIX (J<=I)
02200    C                (OUTPUT,REAL)
02300    C
02400    C      FIRST=.TRUE. IF THIS IS THE FIRST CALL OF
02500    C      THE ROUTINE OR IF THE DISTRIBUTION HAS
02600    C      CHANGED SINCE THE LAST CALL OF THE ROUTINE,
02700    C      OTHERWISE SET TO .FALSE.
02800    C              (INPUT,LOGICAL)
02900    C
03000          DIMENSION H(N),D(N*(N+1)/2),F(N*(N+1)/2),X(N)
03100          LOGICAL FIRST
03200          SAVE N2
03300          IF(N.LT.1)THEN
03400              PRINT*,'SIZE OF VECTOR IS NON POSITIVE'
03500              STOP
03600          END IF
03700          IF(FIRST)THEN
03800              N2=2*N
03900              F(1)=SQRT(D(1))
04000              Y=1.0/F(1)
04100              DO 10 J=2,N
04200    10        F(J)=D(1+J*(J-1)/2)*Y
04300              DO 50 I=2,N
04400              V=D(I*(I-1)/2+I)
04500              DO 20 M=1,I-1
04600    20        V=V-F((M-1)*(N2-M)/2+I)**2
04700              V=SQRT(V)
04800              Y=1.0/V
04900              F((I-1)*(N2-I)/2+I)=V
05000              DO 40 J=I+1,N
05100              V=D(J*(J-1)/2+I)
05200              DO 30 M=1,I-1
05300    30        V=V-F((M-1)*(N2-M)/2+I)*F((M-1)*(N2-M)/2+J)
05400              F((I-1)*(N2-I)/2+J)=V*Y
05500    40        CONTINUE
05600    50        CONTINUE
05700          END IF
```

```
05800          DO 60 I=1,N
05900    60    X(I)=H(I)
06000          DO 80 J=1,N
06100          Y=RND( )
06200          DO 80 I=J,N
06300          X(I)=X(I)+F((J-1)*(N2-J)/2+I)*Y
06400    80    CONTINUE
06500          RETURN
06600          END
```

Appendix 2[†]
Graphical methods for sampling from standardized gamma distributions ($\alpha \geqslant 1$) and the standardized normal distribution

There are occasions when small random samples are rapidly required from the more common distributions (if only for setting examination questions!) without recourse to a computer or even a programmable calculator. Appendices 2.1–2.14 contain figures for the manual generation of variates from standardized gamma distributions with shape parameters $\alpha = 1$, 1.25, 1.5, 1.75, 2, 2.5, 3, 4, 5, 6, 7, 8, 9, and 10, respectively. To generate a variate, a pair of random numbers R_1, R_2 is used to locate the position of a point in the unit square (R_1 representing the abscissa). If the point lies inside the central acceptance region the gamma variate is delivered as

$$X_a = (\alpha - 1) + a(R_2 - b)/R_1,$$

where a and b have been specified for each figure. Otherwise the trial point is rejected and further pairs (R_1, R_2) are tried until acceptance is finally achieved. The method is 'exact', except for the small proportion of prospective points which lie close to the boundary of the acceptance region. In such cases there is a risk of misjudging which side of the boundary the point lies. Given that this manual method is to be used only for small random samples, the quality of the variates produced (as measured by a test of fit) is unlikely to be discernibly different from a corresponding sample of 'exact' variates.

The method is limited to those values of α for which a figure is shown. Approximate variates for intermediate values of α may be drawn by sampling from two figures (parameter values α_1 and α_2 with $\alpha_1 < \alpha < \alpha_2$) with probability $(\alpha_2 - \alpha)/(\alpha_2 - \alpha_1)$ and $(\alpha - \alpha_1)/(\alpha_2 - \alpha_1)$, respectively. For small

---

[†] Appendices 2.1–2.8, 2.9, 2.12, 2.14, and 2.15 are reproduced from Dagpunar (1987), by permission of the author and the Editor of *The Statistician*.

random samples this is usually adequate in view of the comments made in the previous paragraph regarding a test of fit. An extra random number is required to decide whether prospective points should be generated from figure α_1 or α_2.

Appendix 2.15 shows a figure for generating standardized normal deviates. These figures allow variates from other distributions to be drawn. Thus for a beta distribution with α, $\beta \geqslant 1$,

$$X_{\alpha, \beta} = X_\alpha/(X_\alpha + X_\beta),$$

where X_α and X_β are standardized gamma variates. To sample from an F-distribution with n_1 and n_2 degrees of freedom ($n_1 \geqslant 2$, $n_2 \geqslant 2$) deliver

$$F_{n_1, n_2} = n_2 X_{n_1/2}/(n_1 X_{n_2/2}).$$

A chi-squared variate with n ($\geqslant 2$) degrees of freedom is delivered as

$$\chi_n^2 = 2X_{n/2}.$$

Finally, a variate from a t-distribution with n degrees of freedom is delivered as

$$T_n = Z/\sqrt{(2X_{n/2}/n)}$$

where $Z \sim N(0, 1)$ is obtained from Appendix 2.15.

2.1 Gamma ($\alpha = 1$)

$$X = 0.7358 R_2/R_1$$

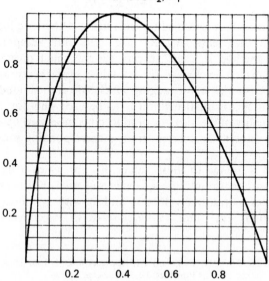

2.2   Gamma ($\alpha = 1.25$)

$$X = 0.25 + 1.1630(R_2 - 0.1624)/R_1$$

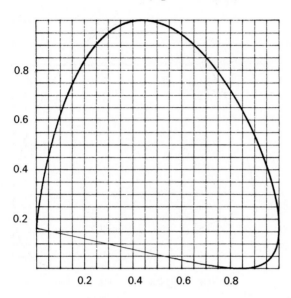

2.3   Gamma ($\alpha = 1.5$)

$$X = 0.5 + 1.4498(R_2 - 0.2262)/R_1$$

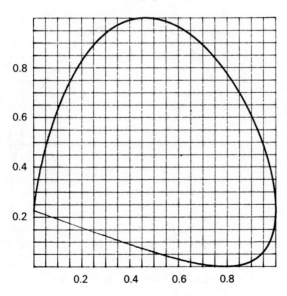

2.4   Gamma ($\alpha = 1.75$)

$$X = 0.75 + 1.6863(R_2 - 0.2635)/R_1$$

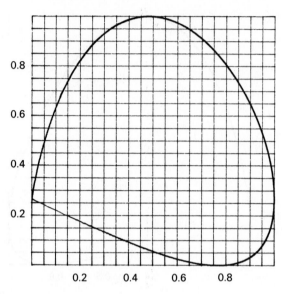

2.5   Gamma ($\alpha = 2$)

$$X = 1 + 1.8929(R_2 - 0.2887)/R_1$$

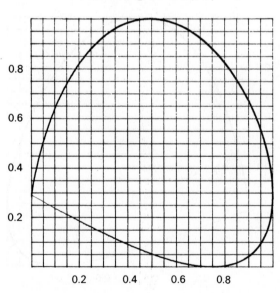

2.6   Gamma ($\alpha = 2.5$)

$$X = 1.5 + 2.2491(R_2 - 0.3216)/R_1$$

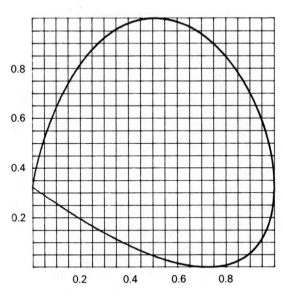

2.7   Gamma ($\alpha = 3$)

$$X = 2 + 2.5559(R_2 - 0.3427)/R_1$$

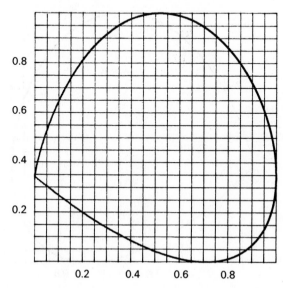

2.8 Gamma ($\alpha = 4$)

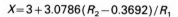

$$X = 3 + 3.0786(R_2 - 0.3692)/R_1$$

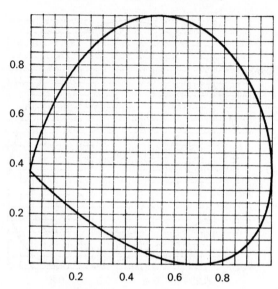

2.9 Gamma ($\alpha = 5$)

$$X = 4 + 3.5245(R_2 - 0.3856)/R_1$$

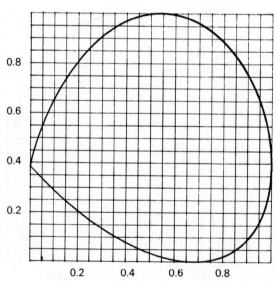

2.10   Gamma ($\alpha = 6$)

$$X = 5 + 3.9199(R_2 - 0.3971)/R_1$$

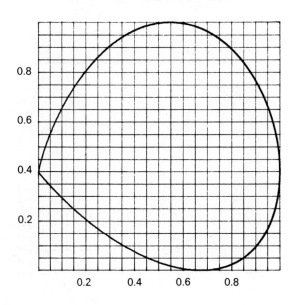

2.11   Gamma ($\alpha = 7$)

$$X = 6 + 4.2790(R_2 - 0.4057)/R_1$$

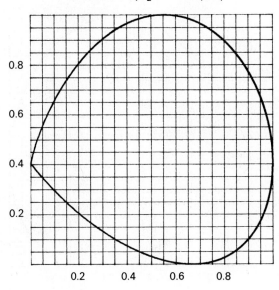

2.12 Gamma ($\alpha = 8$)

$$X = 7 + 4.6101(R_2 - 0.4124)/R_1$$

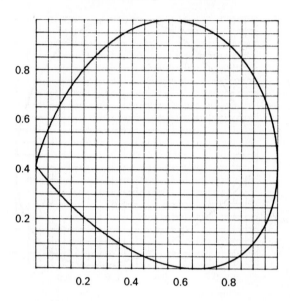

2.13 Gamma ($\alpha = 9$)

$$X = 8 + 4.9190(R_2 - 0.4180)/R_1$$

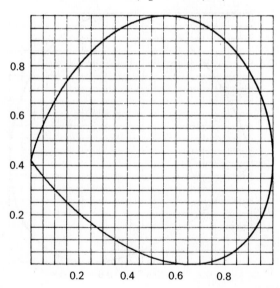

2.14   Gamma ($\alpha=10$)

$$X=9+5.2096(R_2-0.4225)/R_1$$

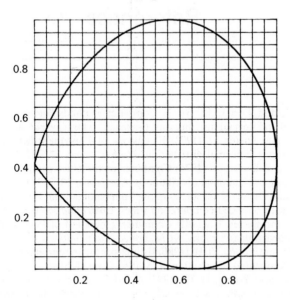

2.15   Normal

$$X=0.8578(2R_2-1)/R_1$$

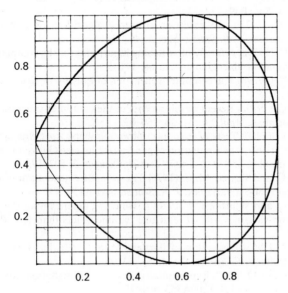

References

Ahrens, J. H. and Dieter, U. (1972) Computer methods for sampling from the exponential and normal distributions. *Commun. Assoc. Comput. Mach.* **15**, 873–82.

Ahrens, J. H. and Dieter, U. (1973) Extensions of Forsythe's method for random sampling from the normal distribution. *Math. Comput.* **27**, 927–37.

Ahrens, J. H. and Dieter, U. (1974) Computer methods for sampling from gamma, beta, Poisson and binomial distributions. *Computing* **12**, 223–46.

Ahrens, J. H. and Dieter, U. (1980) Sampling from binomial and Poisson distributions: a method with bounded computation times. *Computing* **25**, 193–208.

Ahrens, J. H. and Dieter, U. (1982) Computer generation of Poisson deviates from modified normal distributions. *ACM Trans. Math. Software* **8**, 163–79.

Al-Saadi, S. D. and Young, D. H. (1980) Estimators for the correlation coefficient in a bivariate exponential distribution. *J. Statist. Comput. Simul.* **11**, 13–20.

Al-Saadi, S. D. and Young, D. H. (1982) A test for independence in a multivariate exponential distribution with equal correlation coefficients. *J. Statist. Comput. Simul.* **14**, 219–27.

Al-Saadi, S. D., Scrimshaw, D. F., and Young, D. H. (1979) Tests for independence of exponential variables. *J. Statist. Comput. Simul.* **9**, 217–33.

Atkinson, A. C. (1977a) Discussion of Ripley, B. D. (1977). Modelling spatial patterns. *J. R. Statist. Soc.* **B39**, 172–12.

Atkinson, A. C. (1977b) An easily programmed algorithm for generating gamma random variables. *J. R. Statist. Soc.* **A140**, 232–4.

Atkinson, A. C. (1979a) A family of switching algorithms for the computer generation of beta radom variables. *Biometrika* **66**, 141–5.

Atkinson, A. C. (1979b) The computer generation of Poisson random variables. *Appl. Statist.* **28**, 29–35.

Atkinson, A. C. (1979c) Recent developments in the computer generation of Poisson random variables. *Appl. Statist.* **28**, 260–63.

Atkinson, A. C. (1980) Tests of pseudo-random numbers. *Appl. Statist.* **29**, 164–71.

Atkinson, A. C. (1982) The simulation of generalised inverse Gaussian and hyperbolic random variables. *SIAM J. Scient. Statist. Comput.* **3**, 502–17.

Atkinson, A. C. and Pearce, M. C. (1976) The computer generation of beta, gamma, and normal random variables. *J. R. Statist. Soc.* **A139**, 431–61.

Atkinson, A. C. and Whittaker, J. (1976) A switching algorithm for the generation of beta random variables with at least one parameter less than 1. *J. R. Statist. Soc.* **A139**, 462–7.

Barnard, G. A. (1963) Discussion of Bartlett, M. S. (1963). The spectral analysis of point processes. *J. R. Statist. Soc.* **B25**, 264–96.

Barndorff-Nielson, O. (1977) Exponentially decreasing distributions for the logarithm of particle size. *Proc. R. Soc.* **A353**, 401–19.

Barndorff-Nielson, O. (1978) Hyperbolic distributions and distributions on hyperbolae. *Scand. J. Statist.* **5**, 151–7.

Barndorff-Nielson, O. (1979) Models for non-Gaussian variation with applications to turbulence. *Proc. R. Soc.* **A368**, 501–20.

Barr, D. R. and Slezak, N. L. (1972) A comparison of multivariate normal generators. *Commun. Assoc. Comput. Mach.*, **15**, 1048–9.

Bays, C. and Durham, S. D. (1976) Improving a poor random number generator. *ACM Trans. Math. Software* **2**, 59–64.

Bendell, A. and Samson, W. B. (1981) The use of rank order distributions for the estimation of the probabilities of rare events. *Third National Reliability Conference, 29 April–1 May, Birmingham, Reliability 81*. Chamelon Press, London.

Besag, J. and Diggle, P. J. (1977) Simple Monte Carlo tests for spatial pattern. *Appl. Statist.* **26**, 327–33.

Best, D. J. (1978) Letter to the Editor. *Appl. Statist.* **29**, 181.

Best, D. J. and Fisher, N. I. (1979) Efficient simulation of the von Mises distribution. *Appl. Statist.* **28**, 152–7.

Beyer, W. A., Roof, R. B., and Williamson, D. (1971) The lattice structure of multiplicative congruential pseudo-random vectors. *Math. Comput.* **25**, 345–63.

Bhat, U. N. (1972) *Elements of applied stochastic processes.* Wiley, New York.

Birnbaum, Z. W. (1952) Numerical tabulation of the distribution of Kolmogorov's statistic for finite sample size. *J. Am. Statist. Assoc.* **47**, 425–41.

Birtwistle, G.O. (1979) *DEMOS: a system for discrete event modelling on SIMULA*, Macmillan, London.

Birtwistle, G. O., Dahl, O. J., Myhrhaug, B., and Nygaard, K. (1973) *SIMULA BEGIN*. Auerbach, Philadelphia.

Box, G. E. P. and Müller, M. E. (1958) A note on the generation of random normal deviates. *Ann. Math. Statist.* **29**, 610–11.

Brent, R. P. (1974) A Gaussian pseudo random number generator. *Commun. Assoc. Comput. Mach.* **17**, 704–6.

Butcher, J. C. (1961) Random sampling from the normal distribution. *Comput. J.* **3**, 251–3.

Chatfield, C., Ehrenberg, A. S. C., and Goodhardt, G. J. (1966) Progress on a simplified model of stationary purchasing behaviour (with discussion). *J. R. Statist. Soc.* **A129**, 317–67.

Chay, S. C., Fardo, R. D., and Mazumdar, M. (1975). On using the Box–Müller transformation with multiplicative congruential pseudo-random number generators. *Appl. Statist.* **24**, 132–5.

Chen, H.-C. and Asau, Y. (1974) On generating random variates from an empirical distribution. *AIIE Trans.* **6**, 163–6.

Chen, W. C. (1980) On the weak form of Zipf's law. *J. Appl. Prob.* **17**, 611–622.

Cheng, R. C. H. (1977) The generation of gamma variables with non-integral shape parameter. *Appl. Statist.* **26**, 71–6

Cheng, R. C. H. (1978) Generating beta variates with nonintegral shape paramaters. *Commun. Assoc. Comput. Mach.* **21**, 317–22.

Cheng, R. C. H. and Feast, G. M. (1979) Some simple gamma variate generators. *Appl. Statist.* **28**, 290–5.

Cheng, R. C. H. and Feast, G. M. (1980) Gamma variate generators with increased shape parameter range. *Commun. Assoc. Comput. Mach.* **23**, 389–94.

Clementson, A. T. (1982) *Extended control and simulation language.* Cle-Com, Birmingham.

Cochran, W. G. (1954) Some methods for strengthening the common χ^2 tests. *Biometrics* **10**, 417.

Coveyou, R. R. (1960) Serial correlation in the generation of pseudo-random numbers. *J. Assoc. Comput. Mach.* **7**, 72–74.

Coveyou, R. R. and MacPherson, R. D. (1967) Fourier analysis of uniform random number generators. *J. Assoc. Comput. Mach.* **14**, 100–19.

Cox, D. R. and Isham, V. (1980) *Point processes.* Chapman and Hall, London.

Cox, D. R. and Miller, H. D. (1965) *The theory of stochastic processes.* Chapman and Hall, London.

Crane, M. A. and Lemoine, A. J. (1977) *An introduction to the regenerative method for simulation analysis.* Springer, Berlin.

Dagpunar, J. S. (1978) Sampling of variates from a truncated gamma distribution. *J. Statist. Comput. Simul.* **8**, 59–64.

Dagpunar, J. S. (1983) Methods for generating variates from probability distributions. Unpublished Ph.D. thesis, Brunel University.

Dagpunar, J. S. (1987) Nomograms for the manual sampling of random variates from gamma, normal and associated distributions. *The Statistician* **36**, 31–36.

Dagpunar, J. S. (1988). Computer generation of random variates from the tail of *t* and normal distributions. To appear in *Commun. Statist.–Simul. Comput.*

Dieter, U. (1971) Pseudo random numbers: the exact distribution of pairs. *Math. Comput.* **25**, 855–83.

Dieter, U. (1975) How to calculate shortest vectors in a lattice. *Math. Comput.* **29**, 827–33.

Dieter, U. and Ahrens, J. H. (1973) A combinatorial method for the generation of normally distributed random numbers. *Computing* **11**, 137–46.

Downton, F. (1970) Bivariate exponential distributions in reliability theory. *J. R. Statist. Soc.* **B32**, 408–17.

Eilon, S. and Fowkes, T. R. (1973) Sampling procedures for risk simulation. *Opl. Res. Q.* **24**, 241–52.

Ellison, D. and Tunnicliffe Wilson, J. (1984) *How to write simulations on microcomputers.* McGraw-Hill, London.

Fisher, R. A., Corbett, A. S., and Williams, C. B. (1943) The relation between the number of species and the number of individuals in a random sample of an animal population. *J. Animal Ecol.* **12**, 42–57.

Fishman, G. S. (1973) *Concepts and methods of discrete event simulation.* Wiley, New York.

Fishman, G. S. (1976a) Sampling from the gamma distribution on a computer. *Commun. Assoc. Comput. Mach.* **19**, 407–9.

Fishman, G. S. (1976b) Sampling from the Poisson distribution on a computer. *Computing* **17**, 147–56.

Fishman, G. S. (1978) *Principles of discrete event simulation.* Wiley, New York.

Fishman, G. S. (1979) Sampling from the binomial distribution on a computer. *J. Am. Statist. Assoc.* **74**, 418–23.

Forsythe, G. E. (1972) Von Neumann's comparison method for random sampling from the normal and other distributions. *Math. Comput.* **26**, 817–26.

Gebhardt, F. (1967) Generating pseudo-random numbers by Shuffling a Fibonacci sequence. *Math. Comput.* **21**, 708–9.

Gerontidis, I. and Smith, R. L. (1982) Monte Carlo generation of order statistics from general distributions. *Appl. Statist.* **31**, 238–43.

Golder, E. R. (1976a) Algorithm AS98. The spectral test for the evaluation of congruential pseudo-random number generators. *Appl. Statist.* **25**, 173–80.

Golder, E. R. (1976b) Remark ASR18. The spectral test for the evaluation of congruential pseudo-random generators. *Appl. Statist.* **25**, 324.

Gonzalez, T., Sahni, S., and Franta, W. R. (1977) An efficient algorithm for the Kolmogorov–Smirnov and Lilliefors tests. *ACM Trans. Math. Software* **3**, 60–4.

Good, I. J. (1953) The serial test for sampling numbers and other tests for randomness. *Proc. Camb. Phil. Soc.* **49**, 276–284.

Good, I. J. (1957) On the serial test for random sequences. *Ann. Math. Statist.* **28**, 262–4.

Goodman, S. E. and Hedetniemi, S. T. (1977) *Introduction to the design and analysis of algorithms.* McGraw-Hill, Tokyo.

Gordon, G. (1975) *The application of GPSS V to discrete system simulation.* Prentice-Hall, Englewood Cliffs, N.J.

Greenberger, M. (1961) An *a priori* determination of serial correlation in computer generated random numbers. *Math. Comput.* **15**, 383–9.

Greenberger, M. (1965) Methods of randomness. *Commun. Assoc. Comput. Mach.* **8**, 177–9.

Greenwood, A. J. (1974) *A fast generator for gamma distributed random variables. COMPSTAT 1974* (ed. G. Bruckman *et al.*) pp. 19–27. Physica, Vienna.

Greenwood, J. A. (1976) A fast machine-independent long-period generator for 31-bit pseudo random integers. In *Proc. in Computational Statistics* (ed. J. Gordesch and P. Naeve) pp. 30–7. Physica, Vienna.

Gregory, C. G. (1979) A simulation model of a telephone answering service. Unpublished M.Sc. dissertation, Department of Statistics and Operational Research, Brunel University.

Gruenberger, F. and Mark, A. M. (1951) The d^2 test of random digits. *Math. Tables Other Aids Comput.* **5**, 109–10.

Hammersley, J. M. and Handscombe, D. C. (1964) *Monte Carlo methods.* Methuen, London.

Hawkes, A. G. (1972) A bivariate exponential distribution with applications to reliability. *J. R. Statist. Soc.* **B34**, 129–31.

Hill, B. M. (1974) The rank frequency form of Zipf's law. *J. Am. Statist. Assoc.* **69**, 1017–26.

Hill, B. M. and Woodroofe, M. (1975a) On Zipf's law. *J. Appl. Prob.* **12**, 425–434.

Hill, B. M. and Woodroofe, M. (1975b) Stronger forms of Zipf's law. *J. Am. Statist. Assoc.* **70**, 212–19.

Hoaglin, D. (1976) Discussion of Atkinson, A. C. and Pearce, M. C. (1976). The computer generation of beta, gamma and normal random variables. *J. R. Statist. Soc.* **A139**, 431–61.

Hoaglin, D. C. and King, M. L. (1978) Remark ASR24. A remark on algorithm AS98: the spectral test for the evaluation of congruential pseudo-random generators. *Appl. Statist.* **27**, 375–7.

Holst, P. (1979) *Computer Simulation 1951–1976. An Index to the Literature.* Mansell, London.

Hope, A. C. A. (1968) A simplified Monte Carlo significance test procedure. *J. R. Statist. Soc.* **B30**, 582–97.

Hopkins, T. R. (1983) Algorithm AS193. A revised algorithm for the spectral test. *Appl. Statist.* **32**, 328–35.

Hsuan, F. C. (1979) Generating uniform polygonal random pairs. *Appl. Statist.* **28**, 170–2.

Hull, J. C. (1977) Dealing with dependence in risk simulation. *Opl. Res. Q.* **28**, 201–213.

Hull, T. E. and Dobell, A. R. (1962) Random number generators. *SIAM Rev.* **4**, 230–54.

Hurrion, R. D. (1976) The design, use and required facilities of an interactive visual computer simulation language to explore production planning problems. Unpublished Ph.D. thesis, University of London.

Hurrion, R. D. (1978) An investigation of visual interactive simulation methods using the job-shop scheduling problem. *J. Opl. Res. Soc.* **29**, 1085–93.

Inoue, H., Kumahara, H., Yoshizawa, Y., Ichimura, M., and Miyatake, O. (1983) Random numbers generated by a physical device. *Appl. Statist.* **32**, 115–20.

Jöhnk, M. D. (1964) Erzeugung von Betaverteilten und Gammaverteilten Zufallszahlen. *Metrika* **8**, 5–15.

Johnson, N. L. and Kotz, S. (1969) *Distributions in statistics. Discrete distributions.* Wiley, New York.

Johnson, N. L. and Kotz, S. (1970a) *Distributions in statistics. Continuous univariate distributions–1.* Wiley, New York.

Johnson, N. L. and Kotz, S. (1970b) *Distributions in statistics. Continuous univariate distributions—2.* Wiley, New York.

Johnson, N. L. and Kotz, S. (1972) *Distributions in statistics. Continuous multivariate distributions.* Wiley, New York.

Jørgensen, B. (1982) *Statistical properties of the generalised inverse gaussian distribution.* Springer, New York.

Kemp, A. W. (1981) Efficient generation of logarithmically distributed pseudo-random variables. Appl. Statist. **30**, 249–53.

Kemp, C. D. (1982) *Low storage Poisson generators for microcomputers. COMPSTAT 1982*, pp. 145–6. Physica, Verlag, Vienna.

Kemp, C. D. and Loukas, S. (1978) The computer generation of bivariate discrete random variables. *J. R. Statist. Soc.* **A141**, 513–19.

Kendall, D. G. (1974) Pole-seeking Brownian motion and bird navigation. *J. R. Statist. Soc.* **B36**, 365–417.

Kendall, M. G. and Babbington-Smith, B. (1938) Randomness and random sampling numbers. *J. R. Statist. Soc.* **101**, 147–66.

Kendall, M. G. and Babbington-Smith, B. (1939) Second paper on random sampling numbers. *J. R. Statist. Soc.* **6**, (Suppl.) 51–61.

Kinderman, A. J. and Monahan, J. F. (1977) Computer generation of random variables using the ratio of uniform deviates. *ACM Trans. Math. Software* **3**, 257–60.

Kinderman, A. J. and Monahan, J. F. (1980) New methods for generating Student's t and gamma variables. *Computing* **25**, 369–77.

Kinderman, A. J. and Ramage, J. G. (1976) Computer generation of normal random variables. *J. Am. Statist. Assoc.* **71**, 893–6.

Kinderman, A. J., Monahan, J. F., and Ramage, J. G. (1977) Computer methods for sampling from Student's t distribution. *Math. Comput.* 31, 1009–18.

Kiviat, P. J., Villanueva, R., and Markowitz, H. M. (1969) *The SIMSCRIPT II programming language.* Prentice-Hall, Englewood cliffs, N.J.

Kleijnen, J. P. C. (1974) *Statistical techniques in simulation. Part I.* Marcel Dekker, New York.

Kleijnen, J. P. C. (1975) *Statistical techniques in simulation. Part II.* Marcel Dekker, New York.

Knuth, D. E. (1972) *Mathematical analysis of algorithms. Information Processing 71.* North-Holland, Amsterdam, pp. 19–27.

Knuth, D. E. (1977) Notes on generalized Dedekind sums. *Acta Arithmet.* **33**, 297–325.

Knuth, D. E. (1981) *The art of computer programming, Vol. 2, Seminumerical Algorithms* (2nd edn). Addison-Wesley, Reading, MA.

Kotas, J. F. and Lau, H.-S. (1978) On handling dependent random variables in risk analysis. *J. Opl. Res. Soc.* **29**, 1209-17.

Kral, J. (1972) A new additive pseudorandom number generator for extremely short word lengths. *Information Process. Lett.* **1**, 164–7.

Krishnamoorthy, A. S. and Parthasarathy, M. (1951) A multivariate gamma type distribution. *Ann. Math. Statist.* **22**, 549–57.

Kronmal, R. A. and Peterson, A. V. (1978) On the alias method for generating random variables from a discrete distribution. Technical Report no. 17, Department of Biostatistics, University of Washington, Seattle, WA.

Kronmal, R. A. and Peterson, A. V. (1979). On the alias method for generating random variables from a discrete distribution. *Am. Statist.* **33**, 214–18.

Kronmal, R. A. and Peterson, A. V. (1981) A variant of the acceptance–rejection method for computer generation of random variables. *J. Am. Statist. Assoc.* **76**, 446–51.

Law, A. M. and Kelton, W. D. (1982) *Simulation modelling and analysis.* McGraw-Hill, New York.

Levene, H. and Wolfowitz, J. (1944) The covariance matrix of runs up and down. *Ann. Math. Statist.* **15**, 58–69.

Lewis, J. G. and Payne, W. H. (1973) Generalised feedback shift register pseudo-random number algorithm. *J. Assoc. Comput. Mach.* **20**, 456–8.

Lewis, P. A. W. (1972) Recent results in the statistical analysis of univariate point processes. In *Stochastic point processes* (ed. P. A. W. Lewis) pp. 1–54. Wiley, New York.

Lewis, P. A. W. and Shedler, G. S. (1976) Simulation of nonhomogeneous Poisson processes with log linear rate function. *Biometrica* **63**, 501–5.

Lewis, P. A. W. and Shedler, G. S. (1979a). Simulation of Non-homogeneous Poisson Processes by Thinning. Naval Research Logistics Quarterly, 26, 403–413.

Lewis, P. A. W. and Shedler, G. S. (1979b) Simulation of non-homogeneous Poisson processes with degree-two exponential polynomial rate function. *Operations Res.* **27**, 1026–40.

Lindley, D. V. and Scott, W. F. (1984) *New Cambridge elementary statistical Tables.* Cambridge Univesity Press, Cambridge.

Loukas, S. (1984) Simple methods for computer generation of bivariate beta random variables. *J. Statist. Comput. Simul.* **20**, 145–52.

Maclaren, M. D. and Marsaglia, G. (1965) Uniform random number generators. *J. Assoc. Comput. Mach.* **12**, 83–9.

Maclaren, M. D., Marsaglia, G., and Bray, T. A. (1964) A fast procedure for generating exponential random variables. *Commun. Assoc. Comput. Mach.* **7**, 298–300.

Mardia, K. V. (1972) *Statistics of directional data.* Academic Press, London.

Mardia, K. V. (1975) Statistics of directional data. *J. R. Statist. Soc.* **B37**, 349–93.

Marsaglia, G. (1961) Generating exponential random variables. *Ann. Math. Statist.* **32**, 899–900.

Marsaglia, G. (1963) Generating discrete random variables in a computer. *Commun. Assoc. Comput. Mach.* **6**, 37–8.

Marsaglia, G. (1964) Generating a variable from the tail of the normal distribution. *Technometrics* **6**, 101–2.

Marsaglia, G. (1968) Random numbers fall mainly in the planes. *Proc. Natl. Acad. Sci. USA* **61**, 25–8.

Marsaglia, G. (1972) The structure of linear congruential sequences. In *Applications of number theory to numerical analysis* (ed. S. K. Zaremba) pp. 249–85. Academic Press, London.

Marsaglia, G. (1977). The squeeze method for generating gamma variates. *Comput. Math. Applic.* **3**, 321–5.

Marsaglia, G. and Bray, T. A. (1964) A convenient method for generating normal variables. *SIAM Rev.* **6**, 260–264.

Marsaglia, G., Maclaren, M. D., and Bray, T. A. (1964) A fast procedure for generating normal random variables. *Commun. Assoc. Comput. Mach.* **7**, 4–10.

Marshall, A. W. and Olkin, I. (1967) A multivariate exponential distribution. *J. Am. Statist. Assoc.*, **62**, 30–44.

Massey, F. J., Jr (1951) The Kolmogorov–Smirnov test of goodness of fit. *J. Am. Statist. Assoc.* **46**, 68–78.

Mathewson, S. C. (1975) Simulation program generators. *Simulation* **23**, 181–9.

Mathewson, S. C. (1977) *A programming manual for SIMON simulation in FORTRAN.* Imperial College, London.

Meyer, H. A. (1956) *Symposium on Monte Carlo methods*, University of Florida 1954 (ed. H. A. Meyer). Wiley, New York.

Michael, J. R., Schucany, W. R., and Haas, R. W. (1976) Generating random variates using transformations with multiple roots. *Am. Statist.* **30**, 88–90.

Mitrani, I. (1982) *Simulation techniques for discrete event systems.* Cambridge University Press, Cambridge.

Moran, P. A. P. (1947) The random division of an interval. *J. R. Statist. Soc.* **9**, (Suppl.) 92–8.

Morgan, B. J. T. (1984) *Elements of simulation.* Chapman and Hall, London.

Naylor, T. H. (1971) *Computer simulation experiments with models of economic systems.* Wiley, New York.

Neave, H. R. (1973) On Using the Box–Müller transformation with multiplicative congruential pseudo-random number generators. *Appl. Statist.* **22**, 92–7.

Newby, M. J. (1979) The simulation of order statistics from life distributions. *Appl. Statist.* **28**, 298–301.

Newell, G. F. (1968) Queues with time-dependent arrival rates (I, II, III). *J. Appl. Prob.* **5**, 436–51, 479–90, 591–606.

Norman, J. E. and Cannon, L. E. (1972) A computer program for the generation of random variables from any discrete distribution. *J. Statist. Comput. Simul.* **1**, 331–48.

O'Donovan, T. M. (1979) *GPSS simulation made simple.* Wiley, Chichester.

Page, E. (1977) Approximations to the cumulative normal function and its inverse for use on a pocket calculator. *Appl. Statist.* **26**, 75–76.

Parker, J. B. (1972) The use of the Monte Carlo Method for solving large-scale problems in neutronics. *J. R. Statist. Soc.* **A135**, 16–43.

Parker, J. B. (1976) Discussion of Atkinson, A. C. and Pearce, M. C. (1976). The computer generation of beta, gamma and normal random variables. *J. R. Statist. Soc.* **A139**, 431–61.

Parunak, A. (1979). Zipf's graphical analysis of ranked counts. *J. Am. Statist. Assoc.* **74**, 25–30.

Payne, W. H. (1977) Normal random numbers: using machine analysis to choose the best algorithm. *ACM Trans. Math. Software* **3**, 346–58.

Pidd, M. (1984) *Computer simulation in management science.*, Wiley, Chichester.

Pritsker, A. A. (1974) *The GASP IV simulation language.* Wiley, New York.

Rabinowitz, M. and Berenson, M. L. (1974) A comparison of various methods of obtaining random order statistics for Monte-Carlo computations. *Am. Statist.* **28**, 27–9.

Ramberg, J. S. and Tadikamalla, P. R. (1978) On the generation of subsets of order statistics. *J. Statist. Comput. Simul.* **6**, 239–41.

Relles, D. (1972) A simple algorithm for generating binomial random variables when N is large. *J. Am. Statist. Assoc.* **67**, 612–13.

Riordan, J. (1958) *An introduction to combinatorial analysis.* Wiley, New York.

Ripley,.B. D. (1983a) Computer generation of random variables: a tutorial. *Int. Statist. Rev.* **51**, 301–19.

Ripley, B. D. (1983b) The lattice structure of pseudo-random number generators. *Proc. R. Soc.* **A389**, 197–204.

Robertson, I. and Walls, L. A. (1980) Random number generators for the normal and gamma distributions using the ratio of uniforms method. United Kingdom Atomic Energy Authority, Harwell (AERE-R 10032).

Ronning, G. (1977) A simple scheme for generating multivariate gamma distributions with non-negative covariance matix. *Technometrics* **19**, 179–83.

Rowe, J. A. (1942) Mosquito light trap catches from ten American cities, 1940. *Iowa State College J. Sci.* **16**, 487–518.

Rubinovitch, M. (1985) The slow server problem. *J. Appl. Prob.* **22**, 205–13.

Rubinstein, R. Y. (1981) *Simulation and the Monte Carlo method.* Wiley, New York.

Sahai, H. (1979). A supplement to Sowey's bibliography on random number generation and related topics. *J. Statist. Comput. Simul.* **10**, 31–52.

Salfi, R. (1974) A long-period random number generator with application to permutations. In *Compstat 1974: Proc. Computational Statistics* (ed. G. Bruckmann *et al.*) pp. 28–35. Physica, Vienna.

Schmeiser, B. W. (1980) Generation of variates from distribution tails. *Operat. Res.* **28**, 1012–17.

Schmeiser, B. W. and Lal, R. (1980) Squeeze methods for generating gamma variates. *J. Am. Statist. Assoc.* **75**, 679–82.

Schmeiser, B. W. and Shalaby, M. A. (1980) Acceptance/rejection methods for beta variate generation. *J. Am. Statist. Assoc.* **75**, 673–8.

Schrage, L. (1979) A more portable FORTRAN random number generator. *ACM Trans. Math. Software* **5**, 132–8.

Seal, H. L. (1947) A probability distribution of deaths at age x when policies are counted instead of lives. *Skand. Aktuarietidskr.* **30**, 18–43.

Seigerstetter, J. (1974) Discussion of Kendall, D. G. (1974). Pole-seeking Brownian motion and bird navigation. *J. R. Statist. Soc.* **B36**, 411–12.

Shuster, J. (1968). On the inverse Gaussian distribution function. *J. Am. Statist. Assoc.* **63**, 1514–16.

Smith, W. B. and Hocking, R. R. (1972) Algorithm AS53. Wishart variate generator. *Appl. Statist.* **21**, 341–5.

Sowey, E. R. (1972) A chronological and classified bibliography on random number generation and testing. *Int. Statist. Rev.* **40**, 355–71.

Sowey, E. R. (1978) A second classified bibliography on random number generation and testing. *Int. Statist. Rev.* **46**, 89–101.

Stephens, M. A. (1974) EDF statistics for goodness of fit and some comparisons. *J. Am. Statist. Assoc.* **69**, 730–7.

Student, (1908) The probable error of a mean. *Biometrica* **6**, 1–25.

Tadikamalla, P. R. (1978a) Computer generation of gamma random variables. *Commun. Assoc. Comput. Mach.* **21**, 419–22.

Tadikamalla, P. R. (1978b) Computer generation of gamma random vaiables—II. *Commun. Assoc. Comput. Mach.* **21**, 925–8.

Tadikamalla, P. R. and Ramberg, T. S. (1975) An approximate method for generating gamma and other variates. *J. Statist. Comput. Simul.* **3**, 275–82.

Tauseworthe, R. C. (1965) Random numbers generated by linear recurrence modulo two. *Math. Comput.* **19**, 201–9.

Thoman, D. R., Bain, L. J., and Antle, C. E. (1969) Inferences on the parameters of the Weibull distribution. *Technometrics* **11**, 445–60.

Tocher, K. D. (1963) *The art of simulation.* English Universities Press, London.

Toothill, A. (1982) PCW subset. *Personal Computer World*, 133 and 173.

Toothill, J. P. R., Robinson, W. D., and Adams, A. G. (1971) The runs up-and-down performance of Tauseworthe pseudo-random number generators. *J. Assoc. Comput. Mach.* **18**, 381–99.

Toothill, J. P. R., Robinson, W. D., and Eagle, D. J. (1973) An asymptotically random Tauseworthe sequence. *J. Assoc. Comput. Mach.* **20**, 469–81.

Walker, A. J. (1977) An efficient method for generating discrete random variables with general distributions. *ACM Trans. Math. Software* **3**, 253–6.

Wall, D. D. (1960) Fibonacci series modulo *m*. *Am. Math. Monthly* **67**, 525–32.

Wallace, N. D. (1974) Computer generation of gamma random variates with non-integral shape parameters. *Commun. Assoc. Comput. Mach.* **17**, 691–5.

Westlake, W. J. (1967) A uniform random number generator based on the combination of two congruential generators. *J. Assoc. Comput. Mach.* **14**, 337–40.

Wichmann, B. A. and Hill, I. D. (1982) Algorithm AS 183. An efficient and portable pseudo-random number generator. *Appl. Statist.* **31**, 188–90.

Wichman, B. A. and Hill, I. D. (1984). Correction to algorithm AS 183. An efficient and portable pseudo-random number genarator. *Appl. Statist.* **33**, 123.

Wilk, M. B., Gnanadesikan, R., and Huyett, M. J. (1962) Probability plots for the gamma distribution. *Technometrics* **4**, 1–20.

Wilson, E. B. and Hilferty, M. M. (1931) The distribution of chi-square. *Proc. Natl. Acad. Sci. USA* **17**, 684–8.

Withers, S. J. and Hurrion, R. D. (1982) The interactive development of visual simulation models. *J. Opl. Res. Soc.* **33**, 973–5.

Zierler, N. (1969) Primitive trinomials whose degree is a Mersenne exponent. *Inf. Control.* **15**, 67–69.

Zierler, N. and Brillhart, J. (1968) On primitive trinomials (mod 2). *Inf. Control* **13**, 541–54.

Zipf, G. K. (1949) *Human behaviour and the principle of least effort.* Addison-Wesley, Reading, MA.

Index